THE ANTIC ROOT

The Antic Root

SHAKESPEARE, AUTOPOIESIS, AND THE CREATION OF THE WORLD

Christopher Pye

FORDHAM UNIVERSITY PRESS NEW YORK 2026

Fordham University Press also publishes its books in a variety of electronic formats. Some content that appears in print may not be available in electronic books.

Visit us online at www.fordhampress.com.

For EU safety / GPSR concerns: Mare Nostrum Group B.V.,
Doelen 72, 4831 GR Breda, The Netherlands, gpsr@mare-nostrum.co.uk

Library of Congress Cataloging-in-Publication Data available online at https://catalog.loc.gov.

Printed in the United States of America

28 27 26 5 4 3 2 1

First edition

Contents

Preface

The Antic Root: Shakespeare, Autopoiesis, and the Creation of the World explores the relationship between art and world in the early modern era. Prompted by ecocritical concerns in particular, a range of current critical modes, including posthumanism, vitalism, new materialism, and object-oriented ontology, have sought to describe the world absent its human framing—the world "as such," in its own right. What such approaches often fail to acknowledge is the extent to which the claim to know the world without us has always underwritten the human subject as a transcendental observer, as a subject founded on its capacity to compass and thus exceed its own limits. At the same time, even as they risk reinstating the subject, such approaches reduce the world to an object that lies before us, available to our descriptive gaze. The current book argues that to think the world otherwise, to see it beyond that familiar subject-object opposition and thus genuinely beyond human claims to mastery, we need to understand the world neither as an objective nor a subjective phenomenon but as an autopoietic one. Art, and especially the work of Shakespeare, can play a critical role in that task of reimagining, I will suggest.

Adopted from systems theory and especially the work of Niklas Luhmann, autopoiesis describes the means through which any system of meaning at once reflexively constitutes and divides itself, in the process instituting the very distinction between an inside and an outside, an entity and its environment, the self and an objective world. Approached as an autopoietic form, the world appears beyond the familiar, already predetermined opposition of subject and object as a more primordial creative force, as that which eludes even as it constitutes the possibility of any meaningful system. As a process that extends beyond the entities it forms, autopoiesis reveals the inhuman not as a transcendental resource but as the very condition of identity and phenomenal experience, an uncanny element associated with material techne, including the techne of the theater, with animality, and with human disability—intersections manifest, as we will see, in instances that range from a peg-legged interloper representing the force of truth in the historical case of Martin Guerre to the clairvoyant power of the blind in *King Lear* to the dwarfs and insensible creatures that embody the enigmatic grounds of Diego Velázquez's art.

Autopoietics holds particular promise for early modern studies, I will suggest. Though critics have recently sought to find in the era a unified cosmic vision as a counter to the depredations of our post-Cartesian modernity, early modernity was in fact marked by the increasing separation of spheres: the social, the economic, the epistemological, the aesthetic. Autopoiesis as a process helps explain how those autonomous domains manage to engage one another—how they constitute a world at all.

But the mechanism is particularly revealing in relation to the emergence of aesthetic consciousness during the era. As against those critical approaches that have attempted to conceive the world immanently, absent its interpretive framing, autopoiesis as a reflexive formation suggests that world is an irreducibly mediated phenomenon, one that can't be separated from the fact of its framing. In that regard, it is precisely in the register of art, and especially in the works of Shakespeare I'll suggest, that we most closely engage the conditions of world creation. As an autonomous, self-referential domain defined by its power to compass and exceed its own bounds, literature played a critical role in making the enigma of autopoietic origination knowable: thus art's significance in relation to the socius. Precisely in separating from the world, aesthetics as an autopoietic mechanism conjured the world, revealing it in the very groundlessness of its divided and irreducibly mediated condition—as a world always elsewhere. At the same time, extenuating the process in its own self-referential form, the artwork as aesthetically conceived broached the limits of autopoiesis itself as a phenomenally knowable mechanism. Paradoxically, it is within the attenuated register of aesthetics as it was emerging during the era that the most radical prospects for thinking the world beyond our human ken come to view.

A significant dimension of Shakespeare's radical originality derives from his recognition of the intrinsically autopoietic character of literary form and how that fact bears on theater's forming relation to world. Ranging across his entire oeuvre, the book argues that the concept of autopoietic creation offers a promising way of approaching Shakespeare, helping us unpack the logic informing the works' generic distinctions. The constitutive dimensions of mediatory framing authorize an invigorated return to the problem of genre; among other things, the book will forward a new theorization of tragedy, comedy, and romance as inherently autopoietic formations. Engaging the full range of the tragedies, the first three chapters of the book show the way tragic finitude as it was conceived during the era involved not a confrontation with knowable limits, not a scene of recognition at all, but a missed encounter that is also the originary condition of subject and world: the missed encounter, for instance, between a king and his daughter or between the blind and the mad at *King Lear*'s conclusion, figures both, the book argues, for the incognizable grounds of any human rapport. That experience of constitutive difference is associated with the autopoietic formation of the world in *King Lear*, with the foundations of economy and polis in *Timon of Athens*, and with the conditions of being as such in *Anthony and Cleopatra*, a work that in its reiterative and eroticized denouements—no end of ending—undoes tragedy itself as a form.

The notable literary self-consciousness of Shakespeare's comedies, the insistence with which through their epilogues the romantic comedies at once frame and bypass their limits as literary forms, reveals the genre's explicitly autopoietic logic, as well as its intimate relationship to emerging conceptions of aesthetic consciousness. The genre's status as a distinctly social form is directly related to that autopoietic logic: Through its limitlessly self-framing enactments, Shakespearean comedy comprehends its own purely mediated condition, thus fulfilling its distinctive and paradoxical ambition to establish a relation to the fact of relationality itself. Rather than beginning with the familiar opposition between a paternal prohibition and youthful desire, a work such as *Twelfth Night*, acknowledging the indissociability of desire and prohibition, law and its transgression, articulates the prior and far more enigmatic origin of relation as such, which is to say, of the possibility of a social world at all.

Bypassing the boundary of death, romance, as Shakespeare engages it, realizes autopoiesis in its purest form, as a form constituted in the very movement through which it exceeds itself, and, in doing so, exceeds every determinable limit between work and its beyond. Effacing the distinction between origins and ends, the genre realizes its world-forming and world-dissolving ambition through a totalizing aesthetic that extends from the most intimate grounds of sense experience to the limits of the witnessed world, compassing the ambient "sounds and sweet airs" of *The Tempest*, the foundations of sovereignty and law in *The Winter's Tale*, and the origins of nation in *Cymbeline*.

Chapter 5 explores that relation between sense experience and world creation by way of a particular theatrical moment that anticipates the late works: the scene from *As You Like It* in which Rosalind, hearing of Oliver's miraculous conversion in the forest, momentarily faints. Ultimately, the scene evokes not just multiple intersecting transitions between given states—conscious and unconscious, sensate and insensate, doubt and faith, fiction and reality—but the prior reflexive constitution of world, a reiterative missed encounter that evokes the enigma of origination as the wound of affect itself. Contagiously extending beyond the frame of the scene and the play, that affective "cause," an aversive torsion that conjures what it avoids, catches the audience in its implicative toils, revealing the untoward grounds of theater's affective hold. Affect's formative wound—the wound of autopoiesis itself—exposes the intractably sensible and insensible grounds of consciousness, even as it manifests world as what is ever our own and not our own.

Understood as autopoietic formations, Shakespeare's late works intimate the historical shift from art conceived as a mimetic form—as a reflection of an already given world—to art as phenomenal condition, which is to say, art in its relation to the more fundamental and more ambiguous process of world creation. I bear out that claim for Shakespeare's place in the larger transformation of art's status in my final chapter by shifting medium, era, and national context in order to consider the work of the Spanish painter Diego Velázquez. The autopoietic paradox of constitutive division, of opening the world by diverging from the world, is fully manifest in the apparent contradiction at the heart of Velázquez's art. A self-consciously mediated art, an art that flags its autonomous status as art, Velázquez's paintings are equally noted for their capacity to conjure the world in all its sensory immediacy. The two traits are, in fact, indissociable: The movement by which his art at once frames and exceeds its framing enacts the movement by which phenomenal experience is constituted at all, a paradox that aligns Velázquez's famous painterly touch with all the figures of disfigurement that populate his canvases—the dwarfs, the insensate beings, human and otherwise.

In all these instances, the book seeks to reconceive what we mean by world and, in the process, to open alternative ways of imagining community and justice. Understanding world as autopoiesis means conceiving it not as what lies before us but as a constitutively missed encounter that leaves us always already absorbed, from the outset in a state of responding. To think the subject in terms of such a divided condition is to imagine a relation to the other founded not on sympathetic correspondence but on one's noncoincidence with oneself, and to imagine a conception of justice according to which, beyond the sovereign assignation of guilt and innocence, all are marked from the outset by an indebtedness that is the condition of being at all.

A brief note on terminology. The concept of the "missed encounter"—a formulation to which I return at various points and bring to bear on systems analysis—in fact derives from psychoanalytic theory, specifically Jacques Lacan's description of the "encounter forever missed" that lies at the traumatic core of subjectivity, evoking the elliptical causation and incommensurability out of which human consciousness emerges.[1] It's a usage that has threaded through my work over the years, appearing particularly in my earlier, psychoanalytically oriented approaches to early modern literature and culture. In that context, the focus on such a causeless cause—subjectivity as constituted in the movement of its vanishing—served, in part, as a counter to what I took to be the determinism of social-constructionist accounts of subject formation. Ironically, I return to the concept now in the context of a posthumanist critical moment in which the problem of the subject, or at least of the fact of interpretive mediation, is at risk of vanishing altogether, doing so not in order to reclaim the subject but precisely in order to avoid its covert resurrection as an implied transcendental locus of meaning.[2] The usage is in that sense strategic; in the end, the very notion of an "encounter," missed or otherwise, may not be fully adequate to an operation of framing that is neither subjective nor objective. Ultimately, the choice between systems theory and psychoanalysis may be irresolvable, suspended rather than resolved by an undecidable referential drive that at once institutes and troubles autopoiesis as a logical process, a force that is no more inward than it is outward, no more human than inhuman.

A version of the third chapter on *Timon of Athens* and the foundations of the polis appeared in *The Journal of Early Modern Cultural Studies*, and the fifth chapter, "Hurt Feelings: Affect, World, and Time in *As You Like It* and Early Modern Studies," was published in *English Literary History*. A version of the chapter on comedy and romance appears in the *Routledge Companion to Renaissance Literature*, edited by Catherine Bates.

Introduction

The World Elsewhere

Especially as a response to the ecocritical turn, recent work in early modern studies has proposed that the Renaissance era harbors the promise of an integrated cosmological order as a counter to the depredations of capital, empire, and anthropocentric modernity, a "holistic order populated by multiple kinds of being" functioning as "an affirmative counter-discourse of experience, being, and ethical relation."[1] As against the hierarchical order of, say, a Tillyardian World Picture or the abstract uniformity of globalization, the "cosmological aesthetic" recently forwarded by Jane Degenhardt and Henry Turner, for instance, involves an immanent relationality that extends beyond the human subject as an organizing locus, evoking "the fundamentally inter-connected, mutually causal relationship between human subjects and the many substances, forms, ideas, spirits and planetary forces."[2] Though "holistic," as definitionally relational, such a world remains open, a matter of becoming rather than established being—"a complex wholeness of coherence that remains open."[3]

The suggestive approximateness in that summary formulation—what, exactly, is an open wholeness?—may be a built-in feature of any attempt to conceive the world as a world in fully relational terms. Though it speaks of the necessity of thinking the relation between particular and universal, difference and commonality in dialectical terms, Ayesha Ramachandran's account of early modern worldmaking ultimately recasts the problem as a matter of finding a "middle ground" or a "synthesis,"[4] formulations that imply that we already have determinate terms and a measurable ground. For Turner, on the other hand, any possibility of meaning in a relationally conceived universe is a function of contingency and difference: "Meaning—the idea—is...

a relational process, frame dependent and thus always circumstantiated."[5] "Consciousness results from the experience of difference—the experience of experience *as* difference."[6] And yet, if meaning and consciousness are constituted through the movement of difference, wouldn't that militate against any return to "the harmony and perfection of the whole"? Wouldn't meaning and consciousness be by definition what exceed such correspondent "unity"?[7]

For Degenhardt and Turner, the task of managing that enigmatic relation between difference and wholeness falls to art. The world as they conceive it is an explicitly aesthetic phenomenon: The era embodied a "cosmological aesthetic," involving art as "worldmaking" and the world itself as "a total work of art."[8] A made thing, the world is at the same time a "dynamic, living agent."[9] For Turner, art itself amounts to "lived experience," poetry comporting with "world-creation" insofar as it is itself "a part of the actual world," an "occurrence within the stream of experience"[10]—"a fully live, affective, and embodied experience."[11] Mediated and unmediated, constructed and living, even as such a cosmological vision conceives the artwork as an immanent part of the world, it translates the world itself into an artwork, aesthetic experience becoming "the sudden apprehension of the entire world as a total work of art."[12] The problem of constitutive difference, of mediation, would seem to dissolve altogether by virtue of such a formulation, absorbed as it is within the chiasmus that conceives art as experience even as it conceives experience—the world itself—as art.

Echoing as it does some of the more problematic strains of nineteenth- and twentieth-century aesthetic ideology, the reference to the "total work of art" might leave us wondering if we have escaped the terms of modernity at all. In that later era, the era in which it becomes a fully articulated philosophical category, aesthetics was explicitly tasked with negotiating the relation between particular and universal, difference and totality.[13] More often than not, that process involved a temporalizing translation of the fact of potentially illimitable difference into an incorporative experience of memory and return, a matter, as Degenhardt and Turner put it, of restoring the "lost memory" of the "harmony and perfection of the whole."[14] For Turner, that process of return seems capable of bypassing the frame of this "frame-dependent" world altogether, an escape from the taint of interpretive practice in an unmediated access that "goes beyond human time and possesses the force of originary, foundational . . . consequences for the world that it makes possible."[15] One can feel in such moments of immediate, visionary access that, in the process of articulating the promise of a historically distinct conception of world, we may have lost the fact of difference altogether.

Of a piece with a general affirmation of the world as harmonic whole,

Turner's equation of poetry with "lived experience" reflects a broader vitalistic turn in literary and especially early modern studies. For Bruce Smith, that means a full-throated phenomenological or ontological embrace of "the wholeness of being that each human speaker knew before language," and thus the recovery of the promise for a "thoroughly physical, thoroughly grounded" version of theatrical experience.[16] Joseph Campana finds in Spenser's verse "the immediacy, the physicality of affective and bodily experience" and the "vitality that inheres in the living flesh" embodied in "the physicality uniquely available in poetry."[17]

Such claims to a transparently available life force are aligned especially with a desire to recast human experience, beyond the mediating terms of conscious cognition, in more thoroughgoingly relational terms, as an immanent element within a larger web of human and nonhuman relations. Karen Raber enjoins us to "reject the dead end of asking how we know things in the world, and instead consider the body's engagement in the living environment as a sensing, voyaging, responsive, embodied subject/object or quasi-thing."[18] That claim for "embodied cognition," as Steven Mullaney describes it, or of an "embodied and extended model of cognition," as Evelyn Tribble and John Sutton formulate it, has had special purchase for early modern studies insofar as the era would seem to harbor the promise of a conception of the human subject before and beyond the Cartesian binaries that are seen to have vexed modernity.[19]

What troubles such accounts is knowing whether the claims of embodiment are in fact consistent with relational determination. Insofar as the subject is genuinely relationally constituted, it will be a function of the mediatory difference that constitutes the web as a relational web, and in that sense not "irreducible bodily" at all.[20] To speak in terms of the "intersubjective *stuff* of feeling," as a significant strain of affect studies has tended to do, is to traffic in a contradiction.[21] Where there is rapport, there is the force of difference, a fact that tends to intrude itself as soon as one conceives how an immanently conceived affect becomes communicable. The "sympathy one piece of flesh feels for another . . . resonates at the frequency at which poetry also resonates," Campana writes: Where there is resonance there is interval, the mediatory difference—the being affected *by*—through which feeling becomes feelable at all.[22] The point is not that an essential sensible experience is mediated by cognition; it is rather that sense qua sense is a function of the division through which it becomes sensible. "Sense isn't a matter of having or making sense (the world, existence . . .). It's rather the fact that sense apprehends itself."[23]

Expanded to the level of the world, that problem of the relation between immanence and mediatory difference can be understood in terms of the insistence of the function of framing, a function that must be held in view, I'll suggest, not just as a matter of acknowledging the historical contingency of one's view of the world but because the world as such *is* that boundless framing, which is to say, the fact of our entanglement; paradoxically, any chance of conceiving the world beyond our own claims to mastery will depend on retaining rather than evading that fact of complicity. To conceive the world in purely immanent terms, to see it as a relational network of "substances, forms, ideas, spirits" absent interpretive intervention, is to deny the fact of relationality as such. That refusal extends, I would argue, to every account that sets out to articulate the world in purely descriptive terms, as objectively available beyond its appropriation. What goes unacknowledged in such accounts is the extent to which the modern subject has always staked its powers on just such a claim to pass beyond its own limits, to compass a world beyond itself, a transcendental claim that is heightened rather than mitigated by the assumption that it could designate the human as merely one contingent element in its capacious purview.[24] Even as it ensures the supervenient place of the subject, such a posture—a posture prevalent, as we will see, in many posthumanist accounts—reduces the world to an object, the mutually confirming counterterm in the familiar subject-object doublet. In other words, no world at all.

With its reflexive and metadramatic turns, Shakespearean theater is, I will argue, especially attuned to the relation between such entanglements and the constitution of world, especially insofar as that awareness opens the promise of a world beyond us. The uneasiness one feels in response to Jaques's "All the world's a stage" set piece derives from his reducing the world to representation—a sequence of types—the claim to universality belied by the sense that the scene largely amounts to an affirmation of Jaques's own sovereignly mastering gaze, which is also to say, a mirror of his own embittered narcissism. But to assume that the fault in such a view lies in his conceiving the living, "embodied" world as a mere theatrical display ignores the disquieting effect of the speech's conclusion, the moment just as Jaques completes his survey when Orlando enters carrying aged Adam on his back. Coming on the heels of Jaques's reference to "last scene of all," "second childishness and mere oblivion, / Sans teeth, sans eyes, sans taste, sans everything," it's difficult not to see the appearance of the all-but-emblematic figures as a continuation of the typing, to feel that we too have not escaped an implicating assertion of reductive interpretive force.[25]

It's a complex moment. If we are caught out here, ensnared in the scene's effects, it is precisely because we presumed to see beyond the frame of Jaques's constrictive theatricalizing. Such a presumption to see, and thus to stand beyond, the act of absorbed seeing recurs throughout the play and amounts to the very structure by which the subject seeks to constitute itself as autonomous. Think, for instance, of the multiplying claims to theatrical power involved in the scene in which we as unseen seers attend to the lords' description of secretly watching Jaques in his projectively identificatory account of the "weeping" deer; "I love to cope him in these sullen fits," says Duke Senior (2.1.65–67). The danger, of course, is that to presume to such a posture is to find oneself all the more anticipated by and inscribed within a scene constituted from the outset as just such a structure of relational observing. The point where we claim access beyond the theater's frame, where we imagine we've encountered the world as "living," immediate, and "embodied," is merely the point where we are absorbed without knowing it, a necessarily unstable point at which the very distinction between representation and its beyond is at risk of dissolving, and with it, our own status as subjects.

The world as much as the subject is lost in that scene of capture, reduced to an already given object that in fact amounts to little more than the subject's deceptively affirming mirror. The possibility of thinking world beyond that drama of mimetic captivation is intimated by the observation that prompted Jaques's declamation. Duke Senior had not said that all the world is a stage, nor that we must consider the actual world beyond our staging. He says: "This wide and universal theatre / Presents more woeful pageants than the scene / Wherein we play" (2.7.137–39). No end of scenes—no end of framing—because subject and object alike are constituted via that framing. And yet, precisely because it is constitutive, such a movement of framing will always exceed us. World in Heidegger's verb-sense of "worlding," as the originative opening beyond the already given opposition of subject and object, emerges in that ceaseless movement beyond.[26] "A world is never in front of me," Jean-Luc Nancy writes.[27] The ambiguously human and inhuman character of that elsewhere can be glimpsed in the lords' reference to the "antic root" that "peeps out" from the shore of the stream Jaques is described as feeding with his tears, as if, at the far reach of all the embedded observers observed observing, as the constituting difference that is the ground or root of being and consciousness, we are already caught within a gaze neither our own nor another's.[28] Effacing the distinction between seeing and spectacle, the detail intimates a human/inhuman otherness more uncannily alien than any thematically described array of creaturely beings however agentially empowered.

To understand how such a conception of world bears on the problem of difference and totality, we need to understand what it means to think the frame as determinative. The problem of the observer as an interpretive issue is generally conceived in terms of what is taken to be two opposed and comprehensive methodological options. On the one hand, a "realist" perspective, which assumes a transparently available world, the world "as it is," a methodological posture to which a considerable segment of early modern studies seems to be returning in its ambition to open a field of view beyond the human. On the other, an approach that acknowledges the fact of our interpretive implication, that considers the analytical object to be inflected or even determined by the interpretive frame the analyst imposes upon it. Recourse to the former, realist account may be driven in part by the sense that the latter condemns us to a presentist account, thus obviating any possibility of a fully historical analysis.

And yet, such a framework, which again tends to remain within the terms of the object-subject doublet, fails to recognize the deeper stakes of the logic of framing. To see that, it is helpful to recast the problem of world in terms of the logic of autopoiesis, particularly as it has been developed in Niklas Luhmann's systems theory. Autopoiesis by such an account describes the movement through which any system of meaning forms by at once reflexively incorporating and dividing itself, instituting in that process the very distinction between an inside and outside, between autoreference and heteroreference.[29] Rather than involving synthesis or dialectical sublation, difference and totality prove radically coterminal by virtue of a movement of division that is at the same time the condition of self-completion. Autopoietics is not autonomy; it is the process through which the very distinction between autonomy and heteronomy is produced in the first instance. Just as a figure constitutes itself at once in and against a posited ground, so "system is the difference between system and environment," the apparent redundancy of the formulation explained by the temporal recursiveness of the process.[30] The possibility of system involves an act of framing, though not a framing dictated by a subject, human or otherwise. As the mechanism through which any domain of meaning—system or subject—derives, autopoiesis involves a forming event that will for the subject have always already taken place. Origination by such an account becomes not a source to which one might return but a matter of reiteratively enacted difference.

Rather than establishing the world as an object before me, autopoiesis evokes world precisely in that missed origination, as the condition of being always already reactively caught up, a condition of originary responding that Heidegger calls "thrownness."[31] Which is to say, it evokes the world not as substance but as sheer relationality, relationality in advance of any form of objectification. Insofar as a subject or any coherent system is a function of the prior reflexive division from which the possibility of relation derives, world inevitably implies a blind spot, the exclusion and difference from which meaning emerges, the closure that enables openness:[32] We have world at all only insofar as it is other than the one wherein we play. Understanding world as the condition of being rather than as an already determined "space of habitation" or "sum of the actually extant" means conceiving it as something other than a domain of "harmonic inter-relationality," even if we figure those relations as "mutually causal."[33] Such accounts of the world as a network of mutually determinative relations—as a "web of reciprocal causes" or of "sympathetic correspondence"—fail to acknowledge the delimiting force by which such a system of equivalencies constitutes itself as a system at all: An infinite play of relation without closure, however inevitably self-defeating that act of closure might be, would be no system at all.[34]

The openness implied by an autopoietic conception of world should be distinguished from the understanding of "possible worlds" as it has been forwarded by pragmatic philosophy. Against a "correspondent" or mimetic account of truth, Ruth Ronen frames such a pragmatic conception in terms of a model of world as at once discursively constructed and objectively available. "Discourse creates an object that exists in some logical space, allowing us to refer to it and to make true assertions about it."[35] Having shown that fiction cannot be defined in strictly immanent terms—no set of internal rhetorical descriptors can explain why any given stretch of discourse can become fictive simply by being designated so—Ronen argues that the fact of fictitiousness is "context-bound," which for her means a result of a culturally determined set of logical propositions. That contextual determination is based, in turn, she claims, on the presumption of "an intentional action (of world-projecting, of imagining, of belief-suspending)," a claim that in relation to fiction means "assuming the author as a source of control."[36] For Ronen, such a logico-discursive orientation allows the recovery of a referential orientation for literary studies and a conception of the fictive world as the "interface" between immanent and contextual domains: "Fictionality, as a pragmatic property, is wholly context-dependent, but at the same time fictional texts register their autonomy in relation to this very context."[37]

Even aside from the difficulty of knowing what such a context-dependent autonomy might mean, Ronen's account remains problematic. It's hard to retain the fact of referentiality as the determining characteristic of a contextually defined fictiveness when reference bears as much on the self-reference that defines the work as autonomous as it does on extrinsic discursive determinants; as such, the force of reference necessarily precedes because it determines the very distinction between autonomy and context, leaving the grounds of that referential drive unexplained. Moreover, contextualization itself is an illimitable, and thus unstable, procedure, one incapable of guaranteeing the "meaningfully structured states of affairs and propositions" that Ronen takes as the given of fictive worlds.[38]

Rather than involving the "interface" between immanent and extrinsic domains, world understood autopoietically arises at the prior, inevitably unstable point at which such a distinction is constituted, where the opposition between auto- and hetero- reference forms. In that sense, world is the precondition for, rather than the reflection of, any "meaningfully structured" discursive or logical state. Rather than being a function of imaginative projection or intention, aesthetics as autopoietic formation involves a self-conscious enactment of the recursive process that institutes any discursive domain, one that constitutes as much as it is posited by author and audience as intentional agents.

Far closer to such an autopoietic conception is Simon Palfrey's analysis of Shakespeare's "possible worlds," a Leibnizian account that thinks the poet's worlds in recursive and self-generative terms. According to Leibniz's monadic theory, worlds are constituted, Palfrey proposes, not according to extrinsic abstract dictates but immanently, via the potentially infinite ramification of their already implicit relational structures. Each monad, which is to say every world, "is at once uniquely individual and a crystal repetition of the entire universe, past and future, distant and close at hand."[39] For Palfrey, that "fractal" understanding of creation captures the distinctly Shakespearean sense that every moment harbors the prospect of other worlds, including unrealized or foreclosed future worlds. "Every point is a potential subject, alive with the tracery of lives."[40]

Whereas for Leibniz that prospect of an already immanent futurity involves a theological presupposition, the prior working of a divine hand, for Palfrey such a conception is a function of the preformed status of the playworld as an aesthetic entity. While the real world is subject to gaps, accidents, and "haphazard amnesia," playlife "could only be exactly what it is. And yet this perfection, granted by the forethought and professionalism of its coming-to-be, by the fact that these present shapes are also preformed, also entails the thought of future, different repetitions."[41] Preformed yet endlessly relational in its meanings, Shakespeare's world amounts to a crystallized infinitude.

Whereas, strictly speaking, a fractal organization implies an open system, one in which self-difference and self-completion converge, Palfrey's account of the playworld is ultimately a matter of presupposed completeness, an assumption that the play as an aesthetic work has a form and destination given in advance. On the one hand, as fiction, the playworld is irreducibly conditional, a matter of infinite possibilities without the assumption of a prior ground or reference. On the other, as predetermined aesthetic form, those infinite possibilities can be seen to be immanent, prevenient, and replete. "Insofar as a playworld is genuinely *possible*, there is no such thing as a random action, without some claim of prefigured purpose or iterative potential."[42]

That claim for the work as at once conditional and already complete informs one of the central paradoxes of Palfrey's account. While the playworld may be radically contingent and nonreferential, that very condition can be designated with absolute deictic self-certainty. Shakespeare's theater functions "beyond shared games of language or gesture, and [is] instead active or immanent, in some silent or intervallic or virtual dimension that is *here* even as it cannot quite be seen, and is *felt*, even as it cannot securely be known."[43] "Ontology *is* perception; perception *is* expression; expression *is* being."[44] Such referential certainty—the claim that Shakespeare's shadowing world self-evidently and self-sufficiently *is*—is bound up with the substantialist and vitalist cast of Palfrey's understanding of aesthetic discourse. Shakespeare's language "does not so much mediate the things in the world as are vital with possible life": "His forms pulse, like sperm, with unlived or possible lives, implied more than explicated."[45]

The sense of a dimension unavailable to us yet nevertheless intuitively there and phenomenally extant is a telltale indication that we are in the domain of aesthetic ideology. What underwrites that somewhat mysterious form of apprehension is the positing of the playworld as an objectively given, if illimitable, space, the assumption, that is, that the line between theater, here associated with unbounded plenitude, and the real world, aligned with gaps, negation, and accident, is self-evident—a surprising assumption given the persistence with which Shakespeare troubles our confidence that we know where to draw the line on fiction. To relinquish that claim to the givenness of aesthetic form would mean shifting from a monadic to an autopoietic conception of world, extending Palfrey's fractal understanding of meaning from a feature of the playworld to the indeterminate condition of that world as a world. Rather than involving worlds within worlds, which is to say, world as always already given, the aesthetic work takes on its world-creative potential insofar as it will have always already exceeded its own bounds, undoing itself in the movement by which it forms itself. In that sense, rather than being a feature of a distinct and fallen real world, gaps and negation and disjunction lie at the formative heart of the aesthetic work's power of phenomenal creation.[46]

Understanding world as autopoiesis means situating it not *there*, as an objective feature of the work, but in the space between, the indeterminate space through which the very possibility of referential determination is opened. "If we take away Leibniz's pre-established harmony and yet stick fast to his vision of infinite points of substance, we are left with a chaos of polysemy and conation," Palfrey writes: "For all Shakespeare's fissioning multiplicity, his playworlds are never like this. Every point is invested with purpose."[47] Yes and no. If Shakespeare's work is not like this, it is solely by virtue of the necessary, and necessarily failed, force through which they become signifiable, a reiterative force that renews the disjunction through which the work as aesthetic form constitutes itself in the first instance. Again, rather than lying before us, work and world entail us, and it is only by virtue of that ensnarement that we have work and world at all.

To see what it means to conceive the world autopoietically rather than as a matter of monadic self-inclusiveness or relational correspondence, consider the scene from *The Tempest* that Degenhardt and Turner offer as an example of a "holistic" "cosmological aesthetic," one they see as counterposed to Prospero's imperial worldview. Caliban's well-known account of his dreaming "gestures to a richly populated realm of unseen spirits, a wild cosmology that he is not only attuned to but comforted by and at home in," they write.[48]

> The isle is full of noises,
> Sounds and sweet airs that give delight and hurt not.
> Sometimes a thousand twangling instruments
> Will hum about mine ears; and sometime voices,
> That, if I then had wak'd after long sleep,
> Will make me sleep again, and then in dreaming,
> The clouds, methought, would open and show riches
> Ready to drop upon me, that when I wak'd
> I cried to dream again.[49]

In his experience of the "sounds and sweet airs of the island as music that lulls him to sleep," Caliban reveals an "inclination to live with and in harmony with the many unseen forces that surround him and that permeate the natural world," Degenhardt and Turner write.[50] What such an account of cosmic alignment fails to acknowledge is the disjunction and loss that gives the account its force. The liminality of the voices and sounds that "hum about [his] ears," whether from within or without we can't be certain, plays out in an enfolding of dreaming and waking that culminates in a moment of grammatical slippage, a shift from conditional supposition—"That if I then had wak'd"—to indicative assertion—"that when I wak'd." If the scene evokes the possibility of giving oneself over to a world beyond oneself and of coming-to by way of such a yielding, it's by virtue of the self-forgetfulness that enables that absorptive waking. We have subject and world, the scene suggests, to the extent that we have implication, which is to say, by virtue of the division and loss through which self and world as a habitable space first emerge.[51]

Insofar as it is foundational, the scene of failed recognition, of missing the moment of one's own coming-to, is irreducible, a fact evident in the larger drama of returning to oneself played out for characters and audience alike at the work's close. Degenhardt and Turner argue that Caliban's dream of capacious attunement should be read against the forms of exclusionary compassment implied by Prospero's orchestrated denouement, the circle within which he gathers the "spell-stopped" adversaries whose movements he has been dictating from afar, a forced totalization that the authors associate with imperialism. But that spatial trope merely duplicates the audience's spectatorial posture in relation to the play's own scripted, spellbound characters as they are gathered within the theater's "wooden O." We reiterate Prospero's act in our judgment of him, the act of defining our capacious waking cognizance beyond the bounds of theater's framing limits. With that reiteration, we turn the object of our mastery into the figure of our own ongoing reiteratively spellbound absorption. "Transparency is paid for with opacity," Luhmann writes, describing the divisively totalizing logic of autopoiesis;[52] the further out, the further within theater's hold.

That truth was already apparent in the scene we behold, for those willing to see they do not see. In fact, Prospero never demands recognition from his adversaries, nor does he offer forgiveness to them. The demand for such a mutual recognition—tragedy's conventionally presumed end—remains bound within a circuit of debt and pardon, an essentially economic logic of abstract equivalencies. Breaking that closed circuit, which is to say opening the possibility of world beyond its predetermined structuring terms, means conceiving exchange otherwise. Prospero does not address his foes, who, conscience-bound, remain closed off from him. He speaks an apostrophe on forgiveness that amounts to a self-address; he addresses himself as other. We should remember that it is the inhuman spirit, Ariel, who first advises forgiveness, a passage through otherness and vicarity that, though Degenhardt and Turner take it as a sign of the magus's innate inadequacy, should be construed as the very condition of rapport.

Meanwhile, Prospero's foes aren't called to account. Sealed from him—"Not one of them / That yet looks on me, or would know me"—they return only at the point where Prospero changes attire, presenting himself "as I was sometime Milan" (5.1.82–83,86). If they return to him, it is in the form of memory, a passage further within via the "backward and abysm of time" that ultimately means a more radical passage beyond: For them, drowned Prospero is a figure returning from the dead. If all meet at the end, it is insofar as each fails to coincide with themself, a convergence of difference and totality realized as the very condition of connectedness. The play enables a movement beyond the recoils of revenge and conscience alike, beyond the self-affirming, self-destroying economy of equivalency, to the extent that it engages a vision of self as already from the outset alien to itself. While Degenhardt and Turner present what they take to be a dream of harmonic correspondence as a subaltern's resistance to an emerging imperial worldview, it would be more accurate to suggest that it is through an enactment of world-formative disjunction that Caliban counters an imperial claim to an objectified world of transparent correspondence and equivalence.

Understood in historical terms, autopoiesis implies an understanding of the early modern era quite distinct from the unifying vision Degenhardt and Turner find there. If entertainments such as the one at Kenilworth imply a poetics at one with "lived experience," as the authors argue, it's insofar as such courtly displays involve not so much the critical promise of a "new course for the future" as the reactive fantasy of an era increasingly marked by a separation of self-determining spheres, which is to say, increasingly governed by a pervasive autopoietic logic.[53] In the political realm, that condition is apparent in the process through which the social domain, increasingly untethered from a transcendental ground, comes to form itself via what Claude Lefort calls "the enigma of its external/internal articulation," the explicitly autopoietic process through which it at once exceeds and incorporates its own defining limits. "The social can't be dissociated from the problem of its institution," Lefort writes.[54] In the epistemological domain, such recursively divisive logic is implicit in the emergence of perspectival conceptions of knowledge, a "frame-dependent" formation that always posits another observer beyond the frame that makes the fact of such framing knowable, a blind spot as the ineradicable mark of knowingness—thus the "epistemological anxiety" attending the awareness of the world as constructed.[55] In the aesthetic sphere, art constitutes itself not through an ontological merging with "lived experience" but precisely as an autonomous if self-exceeding domain governed by its own laws of development. Constituting itself as a recursive system, art during the era "explode[s] the world as ontology," and in doing so "taught how to observe oneself as observer."[56] For Luhmann, existence as such *is* that fact of reflexive relation: "Being observed exists owing to being observed. It has no existence outside this situation."[57]

To understand art as autopoiesis is neither to see it as the embodiment of the world nor as mere artifice. More paradoxically, it means that precisely by way of its separation art will have constituted itself as complete and phenomenally "lived." The paradox is apparent over the broad course of Shakespeare's oeuvre, where the increasing self-conscious artifice of the late romances coincides with an increasing immersion in the "sounds and sweet airs" of the human and inhuman sensorium. But that contradictory logic was present all along, particularly at the points where the plays encounter the formative limits of their worlds. Edgar concludes his vertiginous account of the view down from Dover's brink, the point at which the brain might "turn and the deficient sight / Topple down headlong," with reference to "the murmuring surge / That on th'unnumbered idle pepple chafes, / Cannot be heard so high."[58] Description exceeds itself in the self-cancelling movement of such a line, evoking the boundary that secures us as that which we have already passed, suggesting that we never knew on what ground we stood. And yet, precisely through the snare of that missed encounter, poetry realizes its full, absorptive power of sensory evocation, the groundless ground of its world-creative power. Just as a figure constitutes itself at once through and against its ground, so art as autopoiesis forms world, but only insofar as it at the same time and ceaselessly divides itself from the world it constitutes.

That association between aestheticization and the conditions of phenomenal experience bears on a crucial irony of art's relation to autopoiesis, one we will return to throughout. Although for Luhmann art is one autopoietic formation among others, aesthetics nevertheless has a distinctive place precisely because it reflects on the process of autopoiesis itself. By absorbing—or attempting to absorb—autopoiesis into its own reflexively self-compassing, self-exceeding form, aesthetics would translate the missed encounter of autopoiesis—its very groundlessness—into a phenomenally knowable form: thus, beyond teaching "how to observe oneself as observer," art's deeper significance in relation to the necessarily inapprehensible "enigma" of social origination. At the same time, precisely in its intensification of the inherent contradictions of self-reference entailed in that claim to autonomy, art as aesthetically conceived reveals the limits of phenomenalization, ultimately troubling autopoiesis itself as a logical mechanism.

For Luhmann, the distinction that inaugurates, or will have always already inaugurated, autopoietic formations involves a mark always with "two sides," an inside and an outside, autoreference and heteroreference. Corresponding to the distinction between signifier and signified, such a two-sided mark is consistent with autopoiesis's own status as a phenomenal, and thus a cognizable, form. "The sign is a form with two sides, and . . . in using it as a sign, we must always move to and operate from the inner side of the form—that is, the side of the signifier."[59] And yet, insofar as it institutes the distinction, autopoiesis necessarily involves a referential drive in excess of, because constitutive of, the auto-hetero distinction, a cut inassimilable to the cognitive formation it enables. Paradoxically, it is in relation to the spiraling contradictions implied by the sheerly self-referential workings of sublimating art that the most primordial conditions of world are intimated—world not just as the scene always elsewhere implied by Luhmann's limitless dialectical reversals of inclusion and exclusion, seeing and blindness, but as a prior, radically undecidable and thus radically implicative force of reference, one that exceeds even as it determines the very possibility of conceiving world as a system.[60]

The insistence of the problem of reference within aesthetic formation opens the larger question of the relation between autopoiesis and history. Even as autopoiesis lets us understand aesthetics in relation to the general multiplication of self-founding spheres during the early modern era, aesthetics lets us recognize the ways in which autopoiesis as a mechanism inevitably exceeds such a developmental account. As against historicist accounts that consider the formation of world in strictly causal terms, the constitutively missed encounter implied by autopoietics aligns world formation precisely with what undoes causality and sequence. Far from abolishing the fact of history, such an undoing of narrative sequence is consistent with the fully dialectical and antitranscendental history articulated by, for example, Walter Benjamin, a historicity in which the present constitutes its meaning through a "tiger's leap" into the past, even as the past realizes itself in its returning.[61] At the same time, to conceive history in nonmimetic terms also means understanding its implicative claims in terms of a more radically indeterminate force of reference, a blind, undialectizable propulsion in which the truth of historical contingency can not be cleanly separated from the interpretive force entailed in making it legible. As at the close of *King Lear*, that conception of history as causeless cause is aligned with a conception of time and of human rapport in which the very distinction between proximity and distance wavers.

The stakes of the distinction between the dialectical and undialectizable dimensions of autopoiesis become apparent in relation to the problem of technology, and especially technological globalism, suggesting a different approach to the phenomenon than that implied by Degenhardt and Turner's opposition between imperialism's rationalized mapping of the world and the fluid reciprocities of a more capacious cosmological aesthetic. Precisely insofar as it is a constitutive mechanism, autopoiesis necessarily exceeds human agency. But that fact in itself doesn't predict its political consequence. In one regard, autopoiesis as supervenient relationality bears on the exorbitant logic of capital in what was for early modernity a coming era of globalism, the expansion of Marx's "general equivalent" to "a general reabsorption of all possible values into [a] value that defines the equivalency, exchangeability, or convertability of all products and forces of production," a leveling process made vivid, as we will see, in *Timon of Athens*.[62]

Within the sphere of modern technological rationality, such a "reign of equivalence" realizes itself, Nancy argues, in the "self-generat[ing], self-complexif[ying]" structure by which "mastery coils back on itself subjecting us to ever-increasing constraints as we try to escape the previous ones, replacing every kind of progress with an aggravation of our condition."[63] With that familiar proliferative spiral of technological control, every "decision becomes such that what it decides goes beyond anything calculable as the effect of some decision."[64] Such recoils, and the reduction to equivalency they entail, are the telltale indicators of the reiterative reversals of an autopoietic structure, where the way out is always the way in and where mastery is the dialectical equivalent of subjection. Because such a regime of equivalence, including the humanistic conception of mutual causation, translates relation ultimately into a relation of the same to the same, it in fact denies relation altogether and, in doing so, denies world.

But the claim to mastery that propels such a reduction to equivalence is based on a refusal of the incommensurability at the heart of autopoiesis's generative force, which is to say, a denial of the fact of finitude itself. The "beating mind" that afflicts Prospero at the moment he is lost to the world he creates is the thought of death. "I am vexed . . . my old brain is troubled. / Be not disturbed with my infirmity . . . A turn or two I'll walk / To still my beating mind" (4.1.158–63).[65] "Every third thought shall be my grave," Prospero says, a fairy-tale association between death and thirds that arises because death amounts to that which exceeds because it opens the mirror doublings through which the self and world consolidate themselves. Prospero takes his self-forgetful absorption in his own constructions as something absolutely dire—"never till this day / Saw I him . . . so distempered," Miranda says—because that reiterative undoing of securing limits is death—it assumes no other form; the thought will *be* my grave (4.1.144–45). According to the logic of autopoiesis, death, the line that should mark the end of fictive substitution and thus secure the subject absolutely, will always assume the form of a reiteratively missed encounter: the echoed syncope of a beating in mind.

Constituted in the movement by which it exceeds itself, autopoiesis here in its specifically aesthetic form is fully realized ultimately in the indistinction between the play and our own responding, a missed encounter that is also the play's generative source. Given the power he displays toward the end in the spellbinding circle he draws around his adversaries, we might well feel apprehension when Prospero turns, still in character, to address us directly at the close.

> Now my charms are all o'erthrown,
> And what strength I have's mine own,
> . . . . .Now 'tis true
> I must be here confined by you,
> Or sent to Naples. Let me not,
> . . . . dwell

In this bare island by your spell
But release me from my bands
With the help of your good hands.
Gentle breath of yours my sails
Must fill, or else my project fails,
Which was to please. . . . . .
. . . . my ending is despair
Unless I be relieved by prayer,
Which pierces so that it assaults
Mercy itself, and frees all faults.
As you from crimes would pardoned be,
Let your indulgence set me free.

(5.1. 319–38)

The audience might clap gently, graciously freeing the magician. Then again, given the perils of captivation figured throughout the drama but especially at the end, an audience would have more tendentious reasons to wish Prospero on his way, to clap with a spell-breaking vengeance, drawing the line on fiction's hold and securing its own knowing place beyond that circle.

But to claim one's power thus, to speed his ship on its way with something more violent than a gentle breath, is, of course, to repeat the tempest with which the play began and thus to be back where one started, all the more bound up within the circuit of vengeance and within the play's inscribing hold.

To release oneself and the play from that double bind—autopoietic creation reduced to dialectical equivalency—would mean giving oneself over with open eyes to fiction's absorptive snare, to linger within the interval at once within and beyond its hold. Here, in the liminal space of our returning to ourselves at *The Tempest*'s close, a space comparable to Caliban's waking dream, that would mean hearing our response—the sound of our own applause—as something returning to us, as that which brings us to even as it brings the play to a close, and thus to hear it as something at once inward and alien like a humming about the ears. For that beating sound was, in fact, already there, reverberating throughout the play, coming both from without, as the sound of the tempest that sets the drama in motion—"And now I pray you, sir, / For still 'tis beating in my mind, your reason / For raising this sea-storm," Miranda says at the start—and from within—"A turn or two I'll walk / To still my beating mind," Prospero says toward the end, when, caught up in his own artful wedding masque, he loses the larger plot of his adversaries' machinations (1.2.175–77; 4.1.158–63). Neither internal nor external, subjective nor objective, the sound that lingers beyond the end figures the self as intimately riven, finite, and noncoincident, at once ahead and behind itself, even as it figures the world as the ellipsis that opens the very distinction between within and beyond, play and audience, self and object.[66] Understood thus, as a reiteratively missed encounter, world creation implies a form of return more originary and more disquieting than the consolidating return to Milan, with its fiction of an already given world.

Uncanny as it is, that experience of intimate dehiscence alone institutes the possibility of a relational world, not one of mutual equivalence between one entity and another but one in which one comes to at all only insofar as one arrives to oneself from elsewhere, from the place of the other. Rather than seeking the promise of our critical futurity via a return to an imagined unity, including the relation-dissolving ontological coincidence of poetry and world, the prospect of a world beyond our globalizing present, beyond the world of unlimited equivalence, lies in a cognizance of the irreducible vicarity that has always been the condition of any world at all.

As an aesthetic category, the very notion of the "Renaissance" is a self-evidently autopoietic conception, if we take the designation to describe a return to the past that also realizes that past in its purest form, as if for the first time.[67] At the same time, the logic of autopoiesis will dictate the formal orientation of the literary analysis that follows in a more specific way, as a focus on genre. My concern with tragedy, comedy, and romance as meaningful terms of analysis may seem critically suspect, engagement with the abstract and historically ambiguous category of genre having yielded in the current critical moment to harder-nosed consideration of material and social causation. And yet, understood in terms of its autopoietic logic, the indeterminate and apparently secondary mediations of generic designation become precisely to the point: According to the missed encounter of autopoiesis, the work is indissociable from its interpretive reception, the "space between" the space of meaning as such. To cite the priority of material context over the process of generic designation is to fail to recognize that it is through genre understood as the field of autopoietic negotiation that the very distinction between text and context, work and world is constituted.

As we will see, generic differences can be understood in terms of the varied ways in which literary works seek to constitute themselves in relation to their necessarily ungraspable autopoietic "grounds." Rather than a confrontation with brute finality, tragedy involves the reiteratively missed encounter with such a consolidating and epistemologically comforting end, a missed event from which world as such, which is to say the world as always elsewhere, is conjured. Understood in terms of the phenomenalizing stakes of autopoietic origination, tragic pathos isn't an affective confrontation with human or inhuman finitude: It enacts finitude in the reiterative failure of such an encounter, as the contingency and incommensurability that constitutes affect as affect and that exposes at the core of sense experience the determining trace of inhuman techne and disability.

Reflexively doubling tragedy's iterative effects, comedy translates—or seeks to translate—constitutive finitude and division into the generative condition of desire as well as the grounds of social relationality as such; comedy's claim as a self-consciously literary and social form is bound up with that aspiration to comprehend mediation itself as originative condition. Extending the self-exceeding movement of aestheticization beyond the boundary of death, romance fulfills aesthetic formation in its most refined and most ambiguous form. Whereas tragedy enacts the law of finitude in its breached enactment of the boundary by which it constitutes itself as a form, romance swallows up the very distinction between work and world, a totalizing ambition that realizes the phenomenalizing vocation of art in its purest form even as it threatens to undo autopoiesis itself as a logical condition. Even as it de-creates world, romance intimates versions of rapport beyond the reciprocal, mediatory terms in which comedy seeks to sustain them.

The chapters that follow will engage autopoiesis and world creation in the early modern era with a focus largely though not exclusively on Shakespeare. Chapter 1 considers the concept of tragedy in relation to recent critical efforts, including ecocritical efforts, to think the world beyond us. Noting the ways in which current variants of posthumanism, object-oriented ontology, and vitalism risk reaffirming the transcendental subject's power to bypass its own limits—to compass a world without it—I suggest that an autopoietic conception of world, with its capacity to unbind the subject-object opposition, registers the inhuman not as a transcendental resource but as that force which at once enables and intimately disappropriates the possibility of phenomenal experience. That autopoietic conception of formative limits has always been most rigorously articulated through tragedy, a form in which the fact of finitude is realized, ultimately, not as a cognizable limit—not as a scene of recognition at all—but as a reiterative ensnarement: Oedipus's fulfilling his fate by fleeing it; Othello's inciting what he most abjures. "Be sure thou prove my love a whore," he declares from the depths of his anguished absorption.

Within the divided spheres of early modernity, tragedy occupies a distinctive place. More than a merely literary phenomenon, the constitutive difference—the law of finitude—that defines tragedy internally also articulates the formative boundary between aesthetic and social domains, that which at once conjoins and disjoins those domains. I explore that claim by considering two inverse instances: on the one hand, the critical intrusion of the aesthetics of tragedy into a historical legal case—the well-known case of Martin Guerre—and, on the other, the apparent absorption of the world beyond within the autopoietic workings of a literary work in *The Tragedy of King Lear*. In both cases, the internal/external limit that defines tragedy as an autopoietic form illuminates a contingency and historicity at the core of vital life, revealing affect's inextricable relation to material techne and power's inextricable relation to disability.

If *King Lear* enacts the limits of the world, the tragedy of *Timon of Athens* explores the autopoietic foundations of the polis. Ultimately that foundation—the "enigma" of the early modern state's emergence as an autonomous form—needs to be understood as a political-aesthetic phenomenon, rather than as a biopolitical one, as Giorgio Agamben would have it. While economy and art are linked in the play by virtue of the mysterious groundlessness of their determining powers, it is autopoiesis as a specifically aesthetic process—the *poetics* of autopoiesis—that allows the fact of those enigmatic grounds of economy and state to become knowable, a convergence between art and the conditions for phenomenal experience highlighted in the play's preoccupation with taste. As the process that converts groundlessness into a speculative form, aesthetics enables the fiction of equivalence that lets it serve as a figure for political and economic foundation.

And yet, because autopoiesis necessarily exceeds the consciousness it enables, even as it serves a consolidating function, aestheticization ultimately translates tragic finitude into a disturbance now at the level of sense experience itself, a contingency associated with a force of history that extends from the play's epitaphic origins to our own efforts to conceive the play in a phenomenally legible form. The chapter explores that untoward element as it manifests in the irreducibly ambiguous trope of burial in the play: Timon's enigmatic discovery of buried gold, the ambiguities of the misanthropic isolate's own burial, and the play itself in its strange continuance well beyond the hero's demise. The consolidation of the polis as unifying ideological form in the play will depend on the management of that exorbitancy, even while the inevitable impossibility of such efforts opens the prospect of thinking socius and world otherwise, beyond the familiar polarities of autonomy and dependence, inclusion and exclusion, misanthropos and philanthropos.

With *Anthony and Cleopatra*, the missed encounter of tragedy extends to an undoing of the very conception of drama as an event-centered form, transforming theater in the process from a mimetic form mirroring an already presumed world to an explicit world-creative enactment. Shifting beyond autopoiesis as a speculative mechanism—a matter of knowing or not knowing, seeing or blindness—the play performs tragedy's missed encounter as the condition of being as such, realizing being not as irreducible substance or vital essence but as the temporal ellipsis of its coming forth, as that which will have already transpired and has yet to occur. The play's often remarked subversiveness as a female-centered and anti-imperial tragedy ultimately plays out as a disruption of the play itself as a mimetic from, a rupture expressed politically in the undoing of the form of ordered narrative history within which Caesar would inscribe the defeated queen—"history according to the victors," as Walter Benjamin describes it—as well as in the play's resistance to our own efforts to establish an identificatory relation to it: Cleopatra evades us, just as she evades Caesar.

In thus dissolving the speculative logic of mimetically conceived history, the play exposes the more primordial grounds of history and reference, history as an inhuman force that undoes even as it enables the speculative logic of narrative historicizing, the latter a fundamentally formalized and atemporal conception embodied in the play in Actium's role as origin and end of imperial historical destiny. Associated with the suspended condition in which the tragedy leaves the audience at the close, a conditional condition that may or may not unbind the snare of tragic autopoiesis, the play's enactment of being's iterative finitude is ultimately associated with a conception of community and world, human and otherwise, that undoes the opposition between exposure and harboring security.

In tragedy, the work engages the law of finitude in the form of its own extrinsic/intrinsic limit. In comedy, that very opposition between work and its beyond is reflectively doubled—or redoubled—gathered into the work as a manifestly autopoietic literary formation, a sublimating process that transforms, or attempts to transform, death's boundary into a generative relation between prohibition and desire. The notable literary self-awareness of Shakespeare's comedies, the insistence with which through their epilogues the romantic comedies at once frame and bypass their limits as literary forms, reveals their knowingly autopoietic character and thus too their intimate relationship to emerging conceptions of aesthetic consciousness. Shakespearean comedy's claim reflexively to comprehend the fact of mediation as such underlies its conceptual ambitiousness. Rather than beginning with the familiar opposition between a paternal prohibition and youthful desire, a work such as *Twelfth Night* articulates the prior and far more enigmatic origin of relation as such, which is to say, with the possibility of a social world at all: A constitutive go-between, Caesario institutes the relational world within which she/he at the same time intervenes.

The inherent exorbitance of autopoietic formation, the work's becoming itself by virtue of reflexively exceeding itself, is fully realized or thematized in the romances, where the aesthetic work's intrinsically totalizing ambition extends beyond the boundaries of death itself. That passage beyond isn't, ultimately, a matter of transcendence; it's a fulfillment of the conditions of autopoietic origination, a fact apparent in one of the romance form's more striking paradoxes: The most artificed of forms, romance is also the most preoccupied with unmediated sensory experience—the "sounds and sweet airs" of the human and inhuman sensorium. Even as they bear out the contradictory grounds of the most intimate dimensions of sense experience, in their self-exceeding scope as autopoietic formations the late romances extend themselves outwardly to incorporate the political domain more explicitly than do the comedies, whether that means staging the origins of sovereignty and law in *The Winter's Tale* or of the nation in *Cymbeline*, an early modern formation whose realization remains, like that of the aesthetic work itself, always still to come.

And yet, precisely in their fulfillment as totalizing aesthetic forms, at the far reach of their subtlizing refinement, the romances imply a threat more dire than tragic pathos: not the encounter between phenomenalization and the law of finitude but an undoing of the very distinction between spirit and matter, living breath and inhuman techne. At that boundary of aestheticization, where autopoiesis as a coherent and knowable mechanism falters, the aesthetic work intimates the alternative political possibilities it harbors: on the one hand, a corporealization of social meaning that risks abolishing the fact of relation altogether and, on the other, a conception of social justice and world beyond every form of aesthetic and political totalization, which is to say, the enactment of being as being-with.

Having looked broadly at Shakespearean comedy and romance as genres to adduce the relation between autopoiesis and phenomenal experience, in chapter 5 I zero in on a particular moment—the scene in *As You Like It* in which Rosalind, having heard of Oliver's conversion in the forest, momentarily faints away—in order to explore the relation between affect and world creation. Ultimately, the scene evokes not just multiple intersecting transitions between given states—conscious and unconscious, sensate and insensate, doubt and faith, fiction and reality—but the prior reflexive constitution of world, a missed encounter that evokes the enigma of origination as the wound of affect itself. Contagiously extending beyond the frame of the scene and the play, that affective "cause," an aversive torsion that conjures what it avoids, catches the audience in its implicative toils, revealing the untoward grounds of theater's affective hold. Affect's formative wound—the wound of autopoiesis itself—exposes the intractably sensible and insensible grounds of consciousness even as it manifests world as what is ever our own and not our own. Ultimately, the scene evokes being not as a matter of embodiment or embeddedness but as a condition of originative responding, a temporally rivening affective response that institutes the world to which it at the same time reacts—being as "thrownness." Undoing sequence, that temporal disjuncture at the same time reveals affective being as inseparable from the very fact of its historicity.

The convergence between sheer artifice and phenomenal immediacy evident in Shakespeare's later works can be read in terms of the larger shift in the interval between the Renaissance and the nineteenth century from an art conceived in mimetic terms—an art of the object—to an art of the perceptual act, that is, to art as a specifically aesthetic formation. In my final chapter, I will engage that transformation by shifting mediums and eras altogether in order to engage the work of Diego Velázquez, specifically his two late large-scale masterpieces: *Las Hilanderas* and *Las Meninas*. The paradox that has always marked the artist's work, its self-consciously mediated condition as art combined with the uncanny immediacy of its sensible effects, obeys a logic that will be familiar to us by now. By understanding these works as specifically autopoietic creations, we can recognize what their inclusive/exclusive logic can tell us about art's formative claims as they extend from the most intimate workings of sense experience and painterly touch to the historical question of Europe's self-creation.

Precisely at the point of its fullest humanistic realization, that aesthetic operation as Velazquez enacts it exposes the limits of human economy and agency in its formative relation to techne, deformity, and animality, an array of concerns we have traced from the prosthetically enhanced interloper in Martin Guerre to the uncanny seeing blindness of Gloucester in *King Lear* to the knowingly insensate creatures populating Arden's forest and the Spanish painter's canvases. At that limit, beyond autopoiesis itself as a cognizable operation, aesthetics opens alternative possibilities of history and world: history as a force of insistence that exceeds even as it determines narrative historicism, including the narrative that lets us neatly locate the origin of aesthetics in developmental terms, and a concept of community absent the logic of sacrifice, guilt, and punishment at the core of Western conceptions of the socius, an understanding of human rapport as missed encounter in which one coincides with the other precisely insofar as one insistently fails to coincide with oneself.

1
The Autopoietics of Tragedy

What place remains for tragedy as a literary or philosophical concern in a critical moment committed to conceiving the world without us? Of course, tragedy is about human finitude, and it is certainly possible to think ecological calamity, for instance, as a "tragic theme of battling against forces beyond our control," with man "made to realize he is one agent among many in nature's elements."[1] Still, the nonhuman turn implied by recent variants of posthumanism, new materialist and vitalist thought, object-oriented ontology, and speculative realism would seem to call for something other than a return to the claims of tragic pathos. In her vitalist celebration of the "inhumanity that connects to the human precisely in the immanence of its bodily materialism," Rosie Braidotti writes: "Death is overrated. The ultimate substratum is after all only another phase in a generative process."[2] Finitude and passion must by such an account be divorced from their age-old relation to the pursuit of meaning. We "need to delink pain from the quest for meaning and move beyond, to the next stage. That is the transformation of negative into positive passions."[3]

And yet metaphysics may not be so easily banished, judging by Braidotti's reliance on familiar supercessionary and oppositional tropes, a matter of the affirming of "positive" as against "negative" passions. Humanism itself may prove hard to shake, if we accept Claire Colebrook's argument that vitalistic, affirmative accounts such as Braidotti's amount to an "ultrahumanism"—a humanism without the human—insofar as they remain committed to "a single life with one single inclination, that of ongoing self-maintenance."[4] Even beyond such affirmations of a singular, self-same, or self-present vitalism, all those variants of "flat ontology" in which the human is no more than one equivalent in the web of relations that make up the world run the risk of reinstating in their perspectival assumptions the subject they seek to dethrone. From what posture does one claim to know the world beyond or the world without us, one in which "the human 'sees itself' in its innate thingness as object" as one among a multitude of "enmeshed beings"[5]? The paradox by which the subject transcendentally affirms itself precisely insofar as it negates itself brings us back to the familiar recuperative dimension of tragedy and to the duplicity to which Derrida pointed long ago in the notion of the "ends" of man, the fact that man has always found his ends in his ending, that humanism has always been a posthumanism.[6]

Calling on us to "reject the dead end of asking how we know things in the world, and instead consider the body's engagement in the living environment as a sensing, voyaging, responsive, embodied subject/object or quasi-thing" hardly resolves the matter.[7] Such claims to mimetic transparency and the resolving force they inevitably entail has always been part and parcel of claims to human mastery. But difficulties remain even in accounts more cognizant of the mediated and speculative condition of any form of apprehension, including of the "sensing, voyaging, responsive" sort. The object-oriented ontology forwarded by Ian Bogost is also committed to engaging "the nature of being rather than the human philosopher's approach to it."[8] But for Bogost, such an object orientation doesn't imply epistemological transparency; at its core, Bogost insists, the object remains opaque—a "black hole." It is the very fact of the irreducibly mediated character of speculative knowledge—its relational character—that allows us to shift beyond an androcentric orientation, Bogost suggests. Things exist "not just for *us* but also *for themselves* and *for one another*."[9] That emphasis on relational determination rules out any reference to an already determined totality. "The world does not exist"—"there is no ur-thing, no container, no vessel, no concept that sits above being such that it can include all aspects of it."[10]

And yet, far from sustaining an object orientation, to describe the world thus, as illimitable relation, would seem to risk undoing the object or "unit"—Bogost's preferred term—altogether in an infinite play of relational differentials. Bogost staves off such a prospect by translating open relationality into a reflective structure. The object may be radically unknowable, and yet, Bogost avers, it contains the world: "A unit encloses a system—an entire universe's worth." Folded back on itself thus, such a mimetic structure enables the contradictory mixture of unity and difference that sustains Bogost's account, an account in which "a unit is isolated and unique" and at the same time "an effect[] of the autonomous actions of interrelating parts of a system."[11] The containment that Bogost rigorously excluded at the level of the world surreptitiously returns at the level of the unit: "'Nutshellery' isn't a bad metaphor for tiny ontology—the condensation of multitudes into dense singularities . . . it implies containment."[12] With that, the metaphysically loaded language of insides and outsides, "molten core[s]" and shells, returns in full force.

The difficulty in such efforts to describe the world purely without us arises from the contradictory imperative to objectify relationality as such, whether that is accomplished at the level of the system as a totality or at the level of the singular unit. To assume that we can arrive at a posthuman account of meaning through an objective description of the world as a network—the recourse of many environmentally oriented accounts—is to fail to ask how an implicitly infinite relational structure constitutes itself as a system in the first instance, how it assumes immanent form. This is where Niklas Luhmann's systems theory holds promise, a theory whose pertinence for posthumanism Cary Wolfe has deftly spelled out.[13] Luhmann describes the reflexive, autopoietic mechanism by which any system of meaning-productive relations constitutes itself by incorporating its own defining limit, exceeding and forming itself in the same recursive movement. Constituting the very distinction between a within and a beyond, inside and outside, such a recursive, autopoietic instance remains unlocatable in time: It will have necessarily already transpired from the moment the system as system appears. Rather than Bogost's worlds containing worlds, such an account implies a ceaseless opening to other worlds in the very movement by which the structure would self-referentially complete itself.

Definitionally limitless and bound up with the very determination of insides and outsides, autopoiesis thus described inevitably includes the interpreter in its inscriptive reach. "There exists no observer-independent, given reality," Luhmann writes;[14] to be environmentally "enmeshed" is to acknowledge that fact of interpretive complicity, and it is only through such a recognition of epistemological ensnarement, rather than through an evasion of it, that we are able to displace the human subject as determining agency. Because autopoiesis necessarily exceeds itself in the movement by which it consolidates, the subject constituted via that self-referential mechanism inevitably occupies a blind spot in relation to the formation of world it inhabits. "Reality is what one does not perceive when one perceives it":[15] Every system implies a vantage from elsewhere, always another world.

Rather than a posthumanism that sets aside finitude in its claim to a world of sheer immanence, a speculative posture that inevitably reinscribes the boundary it refuses, autopoiesis locates the inhuman in a reiterative mechanism that unbinds the distinction between finite and infinite, closed and open, immanent and transcendent, life and death. Instead of a "monolithic but multiply-tiered ontology" or a cumulable plurality of worlds, world thus conceived is that which forms itself solely in its torsional movement beyond itself.[16]

Though tonally it's difficult to imagine any commonality between the cool rigor of Luhmann's systems analysis and, say, *King Lear*, I would like to argue that, historically speaking, the most sustained engagement with autopoiesis as world creation is tragedy. A genre defined by the incorporation of ends, including its own self-constituting end, tragedy's force has always derived from the undecidability of its own reflexively formative condition. What makes tragedy tragic isn't the fact of death but tragedy's iterability, an iterability that makes death's consolidating finitude as insistent and unlocatable as it is at *Lear*'s close. Tragedy involves not the downfall of the good but the reflexive precipitance of the bad out of the good: Oedipus realizes his fate precisely by seeking to escape it; Othello demands the consequence he would most avoid: "Make me to see't ... be sure thou prove my love a whore." "So I will turn her virtue into pitch, / And out of her own goodness make the net / That will enmesh them all," Iago declaims.[17] "Com'st thou . . . from / The world's great snare uncaught," Cleopatra asks Anthony toward the close.[18] That autopoietic dimension, everything that gives tragedy the character of a trap, informs the way in which tragic contingency inscribes itself "inwardly" and inescapably, as the unavoidable condition of fate or, in tragedy's modern iteration, of consciousness—as the "*law* of finitude."[19]

Understood by way of tragedy, environmental "enmeshment"—the human subject's condition as relationally constituted—isn't a matter of its being "embedded" in a given world, natural or otherwise, a conception that maintains the categories of self and world intact as forms capable of encounter, a matter of "contact zone[s] ... where species meet ... and eyes are opened."[20] Instead, understood in the light of tragedy and the temporality of autopoiesis, enmeshment means the ensnarement by which being comes to itself at all, a missed encounter from which self and world alike appear.[21]

Drawing out that conceptual promise means reading tragedy against its formidable recuperative powers, its capacity to translate loss into the measure of an apparently limitless human inwardness. That solidarity between finitude and humanism is, of course, writ large in the Renaissance in the form of the historical pathos by which the era defined itself as a literary formation in a mimetic relation to a lost past.[22] As we will see, what resists, or holds the potential to resist, that recuperative assimilation of difference, what dissolves mimesis into open iterability, is the inescapable relation between tragic autopoiesis and techne, the techne of tragedy's mediated condition as aesthetic form but also techne as the insistence of the material element that at once enables and undoes the logic of tragic recognition: a stone, a feather, a prosthetic limb. Traced to that ambiguously human and inhuman dimension, the point at which human pathos coincides with the powerless power of disability, tragedy ultimately intimates the limits of autopoiesis itself as a systematic, cognizable mechanism. In doing so, tragedy lets us locate the nonhuman not in an objectifiable realm beyond but at the very core of affective life and self-reference and thus at the origin of the sensible world.

Given our own positions as moderns, it may be impossible fully to comprehend what classical tragedy might have meant before the separation of spheres, which is to say, before the emergence of the category of the literary. Still, it's not surprising that, as we will see, from Aristotle to Hegel and beyond, tragedy has always been understood as more than a strictly literary phenomenon, for, engaged with formative limits, tragedy bears on the indeterminate boundary condition of system as such, whether we conceive system in relation to art, law, consciousness, socius, or world. In the context of early modern tragedy, that boundary logic implies not a merging of domains but a generative interference, the inevitable internal/external trace of every autopoietic formation. Even as it can figure the disruptive, inassimilable, yet formative intrusion of the "real" into aesthetic formation—a constituting "irritant," as Luhmann blandly puts it—tragedy can suggest the no less incalculable intrusion of aestheticization, with all its affective indeterminacy, into the core of social formation, a fact that ultimately suggests tragedy's capacity to unsettle even an exactingly logical account such as Luhmann's.

The historicity of tragedy is indissociable from the limits of that autopoietic mechanism, the point where history reveals itself not as "external" ground as opposed to the "internal" space of theater but as a more fundamental and undecidable inscriptive condition. Here, I'll explore that claim for tragedy as a boundary formation through a pair of examples, one—the well-rehearsed case of Martin Guerre—a historical legal case that "resolves" itself in aesthetic terms, the other a tragedy—*King Lear*—that seems to absorb the world beyond into its reiterative limits, undoing itself as a mimetic form in the process.[23] In both instances, such infraction—the inassimilable causeless cause of autopoietic formation—takes us beyond the world as objectively conceived to the more primordial conditions of world creation, a dimension in which living being proves inextricable from material techne, capacity inextricable from disability, and history inextricable from affective force.

1

Martin Guerre's story is by now a familiar one. For literary critics, the case, and Natalie Davis's narrative of it, was brought to prominence by Stephen Greenblatt's 1990 essay "Psychoanalysis and Renaissance Culture," a piece that used the legal saga as a way to raise the question of the historical limits of modern subjectivity. For Greenblatt, problematizing that conception of the subject as autonomous agent meant recasting it as an object within a network of mutually defining relations. "At issue is not Martin Guerre as subject but Martin Guerre as object, the placeholder in a complex system of possessions, kinship bonds, contractual relationships, customary rights, and ethical obligations," the subject amounting to a product of "relations, material objects, and judgments."[24] Greenblatt speaks here as a social constructionist, but it isn't hard to recognize the pertinence of his formulations for posthumanism debates by extending the range of "relations and material objects" beyond the social domain.

But is it any more legitimate to presuppose such a field, however expansively conceived, than it is to presuppose a subject, subject and totalized world having always amounted to mutually reinforcing categories? Troubling such an opposition isn't a matter extending the network's reach further to include an alien and nonhuman cosmos; it means posing the question of how the network of relations constitutes itself as a network in the first instance. A tale about competing stories passed on by the "native" villagers, by the judges nominated to adjudicate them, by prominent contemporaries, by modern historians—Natalie Davis most notably—and storytellers—novelists, filmmakers—and by the scholarly community that presumes to make sense of it all: Paradoxically, it is Martin Guerre's extravagantly mediated status, the sense that it involves no end of retelling, that takes us to the essence and limits of the social, for sociality is that fact of mediation, of constitutive relationality. The "event of the social extend[s]," Bruno Latour writes, right "through to the event of the reading of the text."[25] Ultimately, in the case of Martin Guerre, as in tragedy, the possibility of meaning—the possibility of a socius at all—consists not in the web of relations as already given but in the autopoietic force that constitutes, even as it troubles, the system as a system, a force ultimately aligned in the narrative with the intervention of law.

Greenblatt's very attention to the overdetermined nature of social causation raises the interpretive issue posed by such a networked conception of social meaning. The subject is, by his account, a "placeholder" in a system of "possessions, kinship bonds, contractual relationships, customary rights ... ethical obligations ... relations, material objects, and judgments" sustained through "a network of lived and narrated stories, practices, strategies, representations, fantasies, negotiations, and exchanges that, along with the surviving aural, tactile, and visual traces, fashion our experience of the past, of others, and of ourselves."[26] Even as it functions as a signifier for historical responsibility, such an incantatory proliferation of associative terms produces the distinct impression that such a listing could go on indefinitely. Insofar as meaning in such a "complex system" is, as Greenblatt claims, truly relational, which is to say differentially determined and thus by definition potentially infinite, what allows the network to be legible *as* a network at all?

Such a conception of relational meaning implies the need for a limit structure, a fact that becomes apparent at what I take to be an important moment in the essay. Having argued at length that "it is important to characterize the case of Martin Guerre as a *story*" in order to explain how local events transmute into larger history, Greenblatt concludes: "And conversely, the trial and execution of Arnaud du Tilh enables us to understand aspects of the social significance of these literary fantasies that would otherwise remain obscure."[27] Even as we must consider history as a matter of stories, the force of law, in this case a fatal force, lets us recognize the significance of these stories in good part by letting us recognize them *as* "literary fantasies." Law's supposed limit, here aligned with the apparently fixed boundary of death, lets us know fiction as fiction, mediation as mediation, thus securing reference and any possibility of coherent historical interpretation. In that regard, such a guarantee equally underwrites Greenblatt's own confident assumption that for the theatergoer the distinction between fictive and real remains always reliably intact.[28]

Even so, the status of such a boundary remains notably ambiguous in this case. For if, as Greenblatt makes clear, nothing is more theatrical in the case of Martin Guerre than its truth-revealing courtroom denouement, what assures the stabilizing opposition between fiction and law, in Greenblatt's account a chiasmic relation that keeps the two "converse" terms distinct even if in play? For much depends on that boundary, including, I'll argue, the possibility of the world as a legible phenomenon. History converges indeterminately with tragedy, "life" with "art," at the enigmatic point of autopoietic self-formation, the infinitely finite point where world, human and otherwise, emerges as a domain of phenomenal meaning.

The story of the "sham husband" who comes to "stand in the place" of another, taking half the village with him in his deception, gives especially vivid narrative form to the claim that the subject is a function of a network of relations, that is, a function of substitution and displacement. It's not just the imposter Arnaud who is defined thus. By Davis's account, Martin himself—the displaced husband—had fled in the first place because he felt himself to be something of a stand-in. Made insecure by his "precarious sexuality," matched with "a wife he may have been afraid of," Martin abandons the patrimonial scene altogether.[29] At the same time, the patriarchal family itself could be said to be in the wrong place, transplanted Basques trying to fit into their adopted Languedoc home. Meanwhile, it can be hard to say if the impostor should really be considered false at all. Everything attests, Davis tells us, to Arnaud "having fallen in love with the wife for whom he had rehearsed and her having become deeply attached to the husband who had taken her by surprise" (44). "Newcomer!" he declares at the arrival of the "real" Martin Guerre, and nothing assures us that he isn't speaking in earnest (84). The possibility that "*Pansette*," the stomach, as he was called, has swallowed everything including himself into his astonishingly verisimilar mimesis or that, like any theatergoer, he has and hasn't given himself over to the imposture has the potential to make history illegible insofar as it extends the reach of the "fictive" indefinitely.

Such a prospect of omnivorous fictiveness casts a different light on the function of usurpation in the tale of Martin Guerre, suggesting that the founding event of the narrative isn't Arnaud's usurping arrival but Martin's prior autopoietically self-defining self-usurpation. For Martin doesn't so much flee the family as engineer his own banishment. Uncomfortable in his role as married heir, Davis writes, "Martin 'stole' a small quantity of grain from his father … Martin Guerre had now placed himself in an impossible situation. 'For fear of the severity of his father,' he left—he left his patrimony, his parents, his son, and his wife—and not a word was heard from him for many years" (24). "Stole" appears here in quotation marks because it's a theft from within, the boy's absconding with the father's "seed" having the earmarks of oedipal ritual, a possibility intimated when Davis continues with her own speculation: "It would be interesting to know whether Martin Guerre retraced his father's journey of two decades earlier and visited the Labourd" (24). Indeed, Martin confidently assumes his identity—his *propre persona*—for the first time only at the point of his return, as something lost and reclaimed.

That logic of formative rupture is equally apparent at the level of the social network. "Just as Bertrande, now mother to a son, was establishing herself on a new footing with her mother-in-law, Martin Guerre disappeared without a trace. This was a catastrophe. Even for peasants who enjoyed a good gossip, the unexpected disappearance of an important villager was troubling, leaving an anomalous gap among the young married couples" (32). A disastrous rent in the social fabric, at the same time it is just such an "anomalous gap" that opens, or will have always already opened, the movement of substitution that underwrites the possibility of meaning and sociality as such.

Precisely as autopoietic "ground," that recursively divisive event exceeds its bounds, extending itself to include the process by which it is made legible. Chapter 1 ends with a dramatized account of the ill-fated couple's marriage night: "Into their room at midnight burst the young revelers, led by Catherine Boeri, a relative of the curate of Artigat. She was carrying their 'resveil.'

Heavily seasoned with herbs and spices, the drink would ensure the newlyweds ardent mating and a fertile marriage" (18). After the page break at this point the next chapter begins: "Nothing happened in Bertrande's bed that night, it seemed, neither that night nor for eight years after" (19). A familiar rhetorical ploy—the narrative historian's equivalent of the cliffhanger or perhaps the dying fall—the moment nevertheless obeys the duplicitous logic of the "anomalous gap," here in relation to the conditions of narrative meaning. "Nothing happened" is of course the purest calamity for historical narration: nothing there. Then again, it's such gaps and suspensions that propel the narrative as narrative—that constitute it as meaningful at all.

That interpretive effect isn't secondary; it is indissociable from the events' status as historical account. Describing poor young Martin under the hand of a severe father and surrounded only by sisters—"pisseuses"—Davis speculates, "Then, when his penis had barely begun to grow behind his codpiece, another girl came into his life, Bertrande de Rols" (20). Of course, much is indeed staked on what is or isn't behind the "anomalous gap" of the boy's codpiece—the possibility of genealogical continuance and thus of genealogical history. At the same time, it is just such withholdings that open the subjectifying space of speculation. An absolutely replete narrative, one as seamless and self-complete as Pansette's consuming inventions, would be as annulling as an empty one: no narrative at all.[30] In that sense, the subject is a function of such a gap, of that which at once falls out of and institutes the space of socius and narrative alike.

Such moments realize the implicative and reiterative character of the "network," that which gives it its interpellative force and thus makes it social at all. What is withheld opens the difference and distance that ensures cognitive mastery. And yet it is precisely that constituting distance that inscribes, our voyeuristic investment in what does or doesn't take place in the marital bed or behind the codpiece making us in this instance the equivalent of the prurient villagers whose intruding gaze prompts the flight that set the story in motion: thus the autopoietic character of the social event, its reiterative expansion by way of an undoing of the distinction between resistance and accession, outside and inside, the event and its observation.

That prospect of evidential groundlessness underlies Davis's moments of hermeneutic extravagance as well as the redundancy of those instances of interpretive force. "The obstinate and honorable Bertrande does not seem a woman so easily fooled, not even by a charmer like Pansette. By the time she had received him in her bed, she must have realized the difference; as any wife of Artigat would have agreed, there is no mistaking 'the touch of the man on the woman'" (44). Mediating through a village aphorism the truth of unmediated touch—that which establishes the proof of a stabilizing difference—Davis's identificatory or transferential relation to Bertrande fills in the critical gap in relational knowledge that underlies the history by translating Bertrande into what Lacan terms "the subject supposed to know": Silent as she is, she if no one else must have been aware.[31] And it is just that grounding assertion of epistemological clarity and mastery that places Davis in the position of all those—husbands, villagers, justices—who, as she has noted, have coercively presumed to speak in the place of the woman. With that, Davis situates herself firmly within the narrative's apparently endless chain of substitutions.[32]

It is in relation to such a possibility of limitless substitution that we can understand the importance of the trial scene, with its enlistment of the authority of law, a law underwritten in this instance by death's finitude and involving an act of judgment that is as much ours as the characters'. But what gives the court's judgment the force of truth involves something other than law understood in its limited empirical sense—the justices were, the lead adjudicator Jean de Coras confesses, on the verge of a mistaken verdict. It is instead what contemporaries described as the "greatest mystery" of the case, the true Martin's arrival in the court at the moment judgment is about to be pronounced. Truth bearing, it is also the drama's most flamboyantly theatrical moment.

Indeed, it's a moment best grasped in aesthetic terms. *The Return of Martin Guerre* observes the principles of Aristotelian tragedy quite precisely: reversal of fortune, anagnorisis, and, perhaps most importantly, the affective force of coincidence.

> Tragedy is an imitation not only of a complete action but of events inspiring pity or fear. Such an effect is best produced when the events come on us by surprise, and the effect is heightened when, at the same time, they follow as cause and effect. The tragic wonder will then be greater than if they happened of themselves or by accident, for even coincidences are most striking when they have an air of design.[33]

That the denouement of a legal drama should assume the form of Aristotelian tragedy may be less surprising than it seems. For Aristotle's innovation—the turn that gave his account of tragedy such force in the literary tradition—was his claim for the commonality between forensic and poetic logic. "At the very foundation of the rapport between tragedy and law in the Aristotelian tradition is Aristotle's extension of the fictional method to include both legal and poetic fiction."[34] While for Aristotle the logic of probability governing poetic works finds a parallel in the introduction of those hypothetic legal fictions that allow considerations of context and motivation, opening the rigor of law to broader concerns of equity, that same merging of poetic and forensic concerns, Lorna

Hutson has argued, lies behind the development of early modern drama as a form based on matters of causality and characterological intention.[35] Mitigating "the rigid generality of the law," that correspondence between poetics and law informs, Kathy Eden argues, a reconciliation between universal and particular that amounts to the very condition of judgment. The evidential orientation of such a forensic poetic, where the fictive image is aligned with causality and "palpable proof," assimilates sense datum to logical argument, "percept" to "concept," thus "enabling the mind to think, judge, will, and eventually to act."[36]

Insofar as it enables such a reconciliation of particular and universal in the formation of judgment, Aristotle's version of tragedy, thus conceived, can feel surprisingly of a piece with Hegel's. That a historical account should take on in its resolving moments the terms of a classical tragedy merely bears out, one could argue, the conjunction between tragedy, law, and history familiar from the Hegelian tradition. It's not difficult to see in the structure of loss and authenticating return that defines the Guerre narrative the outlines of the Hegelian speculative dialectic, the process by which the human subject instates itself through a reflective encounter with its own finite bounds; if for Aristotle causally logical ordering mitigates the rigors of law, for Hegel it's the rigor of law as finality that constitutes the possibility of speculative reason. Tragedy has a privileged place in that dialectic formation for Hegel insofar as it articulates the founding relation between particular and universal, concrete and abstract, subject and society, structured around death's defining limit, death serving at once as "the great equalizer" and the most radically singular of events.[37] Tragedy in that sense confirms the subject's status as a definitionally contingent, historical, being even as it stabilizes the play of difference via a movement of individual and social subsumption, the gathering in of division as self-refection. Recognition, reversal, and the punitive force of law converge in such a dialectic formation, a self-divisively punitive, guilt-inducing moment insofar as consciousness emerges as the defining internalization of what it stakes itself against.[38] Tragedy, according to such an account, amounts to the "birth-trauma of moral conscience."[39]

Still, some element of tragedy—perhaps the element that makes tragedy tragedy—is not accounted for in such descriptions, whether the presumed correspondence between particular and universal in those accounts is seen to enable the "palpable image" to coincide with an imperative "not simply of acting . . . but of acting in the right frame of mind," as Eden claims of Hamlet, or a subjectifying internalization of law, as Hegel would have it. That unsubsumable element can be adduced from what amounts to the most striking point of correspondence between Aristotle's account and the scene as it unfolds in

Artigat: the role Aristotle allots to coincidence. That Aristotle, who has placed himself firmly on the side of probability and orderly progression, should affirm the importance of surprise is not in itself a surprise. To be caught out by a design is to affirm that design as something supervenient, beyond contingent machination, something that befalls us like "an act of providence," Davis remarks, speaking of the dramatic moment (102). Some such logic might be seen to underlie the most striking paradox in Aristotle's analysis: the fact that the ordering logic of probability, and thus the very concept of intention, emerges, "like the peripeteia itself . . . contrary to intention and expectation."[40]

And yet Aristotle isn't just saying that a surprise that follows as effect from cause satisfies the pleasure of design. He says it heightens the tragic wonder. Whereas an accidental occurrence, one that strikes out of the blue, leaves the categories of order and chance intact, a coincidence has the character of the uncanny because it suspends the distinction between design and randomness, necessity and contingency, just as Martin's startlingly punctual return leaves us suspended between truth confirmed and sheer contrivance. If tragic wonder has punctual, law-like force, the power to close the space of fictions from within the space of fictions—the ambition of any autopoietic formation—it's because it entails a cognitive impasse, a hitch in the structure it enables.

The wonder-inducing ambiguities extend beyond the timing of Martin's return. Particularly given that there had been rumors of a war injury at various points, the fact that Martin makes his appearance on a wooden leg is confirming: No one could fake such evidence. Then again, that the proof of the true man's authenticity takes the form of an "anomalous gap" and a substitutive artifice might give us pause. Or, to reverse the reversal, it may bear out the truth of the man but as one who was never fully there.

That relation between anagnorisis and artifice is present in Aristotle as well. Having remarked that "even coincidences are most striking when they have an air of design," Aristotle continues: "We may instance the statue of Mitys at Argos that fell upon his murderer while he was a spectator at a festival, and killed him. Such events seem not to be due to mere chance. Plots therefore constructed on these principles are necessarily the best."[41] The sense of a reverting punishment, of something returning from the past as if from the dead, resonates with the early modern narrative of a revenant's return against those who may have let themselves forget, either by virtue of love or, perhaps in our case, by virtue of readerly absorption. Such punitiveness is in both cases further aligned with a crossing over between life and inanimacy: a sentient statue, a wooden prosthesis as ground of human truth.

To make sense of these dimensions of the scenes, we should note another riddle. Why should Aristotle's single instance of the affective logic of coincidence assume the form of a theatrical scene, the recoiling revenge of a simulacrum against the murderer in his role as "spectator at a festival"? Whether we're speaking of the statue's revenge or the meta- or hypertheatrical return of the prosthetically enhanced husband, such examples suggest that the limit of apparently infinite substitutability—the arresting point of real and poetic justice and the condition of referential truth—arises from within rather than beyond the space of artifice, from a fatality inherent to the logic of substitution itself, the poetically just revenge of the simulacrum corresponding with a fatal recursiveness in the image.

In what sense does art punish? Certainly Martin's return can have the force of a return of the repressed insofar as we, like Pansette, have given ourselves over to an invented narrative. But that humanization of the moment fails to recognize that the threat it entails involves precisely a dissolution of the distinction between the human and the inhuman. If socius and subject alike are a function of the network of differences, then the limit of substitutability, the founding limit of the totality of exchanges, will be the point where the effort to comprehend that difference eludes itself, where, confirming ourselves as existent, we reflexively encounter ourselves as what is already inassimilably alien—a sign and a remnant or revenant. That will remain an incognizable limit insofar as it assumes the form of a missed encounter, and one in which death's securing limit proves at once absolute and reiterative, leaving us always at once within and beyond the formative scene.

The affective force of such a moment of recognition, or nonrecognition, is a function of its temporality. Davis aptly connects Martin's reappearance with the familiar Horatian aphorism "punishment comes on a limping leg but nonetheless catches up with even the fleetest criminal" and reproduces Otto Vaenius's emblem of the saying, one in which punishment strides on a wooden leg, the recoils of vengeance conveyed in the coiled heads of the snakes with which he punishes and in the chiming postures of perpetrator and punishing victim (122). If the lagging step of punishment nevertheless catches its object, it is by virtue of the very irreducibility of that delay. Bypassing itself in the movement by which it comes to, the subject is constituted in the form of a rift or syncope, a division that inscribes itself with the alien force of law—the enigmatically internal and external kernel of autopoietic formation. Law assumes its ineluctability as a consequence of the indissociability of response and threat implied by such a divided coming-to, a response that constitutes what it at the same time flees. Law in that sense is indissociable from tragic pathos insofar as we conceive pathos as an inherently divided affect, one in which the opposed terms of Aristotelian catharsis—pity and fear, identification and recoil—turn out to be indistinguishable. Insofar as those terms—identification and difference, attraction and aversion—constitute the minimal elements of any social formation, tragic pathos enacts the unstable limits of the social as a coherent system of exchange, including, ultimately, even those forms of limitlessly reversible exchanges between insides and outsides, auto and hetero, implied by systems theory.

The affective correlate of the divisive process of autopoietic formation, tragic pathos should be seen to *befall* the subject, arriving from elsewhere, as something alien. In that sense, pathos itself is inhuman. The lagging, precipitous limp of the peg-leg pursuer conveys affect itself as a disability, a paradoxical capacity for incapacity, for not being able, suggesting passion's defining association with passivity. Implying a condition of receptiveness in advance of subject or object, thus dismantling the opposition between transitivity and activity, tragic pathos unbinds the logic of agency as such.[42] As against those vitalist and new materialist accounts that seek to pass beyond anthropocentrism by attributing agency to the things of the world, tragedy locates the fact of finitude beyond the active-passive doublet, implying receptivity and exposure as originative, the primordial condition of being.

2

Aristotle's account of tragic wonder as a form of return that troubles causal logic touches on an ambiguity at the core of the aesthetic function. On the one hand, as a "merely" literary or seemingly nonreferential concern, aesthetics is bound up with the self-founding movement by which it would reflexively comprehend its own limits and thus claim the knowing unknowingness—the wonder—distinctive to it as an epistemological condition; it wouldn't be wrong, according to a certain form of historical accounting, to associate Shakespeare and Montaigne with the coming to prominence of that aesthetic vocation. In its broadest acceptation, aesthetics is indissociable from the larger question of how one makes mediacy "as such"—the groundless ground of meaning—phenomenally available; in that regard, all those accounts that would descriptively comprehend the world as a purely relational network amount to essentially aesthetic projects.

On the other hand, precisely as a self-referential formation, aestheticization inevitably exceeds itself as a speculative mechanism, reiterating the divisive force that enables such a process of phenomenalization:[43] Recall the way Davis's affirmation of touch as the mimetic confirmation of a founding difference implicates her all the more thoroughly in the blind vagaries of substitution. As an aesthetic matter, founding coincidence always wavers between correspondence and sheer, cognitively inassimilable randomness. If for Luhmann autopoiesis always entails a reference beyond the structure it enables, aestheticization, precisely in its claim to self-referential autonomy, exposes a contradiction within the referential drive itself, the necessity not of referring to the self or to the other but of a initiatory relation to nonrelation as the condition of any relation whatsoever.[44] Understood in these terms, rather than being the already assumed medium within which autopoietic forms unfold, history takes effect as the force of contingency that at once disturbs and enables phenomenal cognizance as such, including the cognizance of autopoietics as a systematic form. The relation between aesthetics and that fact of historical contingency is realized ultimately in the form of a performative reiteration, a stagger-step in the argument as truth claim, including the claim for a correspondence between history and the artwork. The dislocating force necessarily involved in any assertion of autopoietic grounds is what ultimately makes Luhmann's account of autopoiesis as a self-referential procedure a political rather than a descriptively sociological or analytical undertaking.

Indeed, the "truth" of autopoietic reference, including the conjunction between disability, inhuman affectivity, and the world formation it entails, may be less apparent *there*, in the galvanizing wonder of Guerre's return, than it is in the displacements of its interpretive afterlife. Among those in attendance at the trial that day was Montaigne, who recounts the judgment's impact in his essay "*Des boyteaux*," "Of Cripples." It's an important essay, articulating in a concise way the larger epistemological stakes of his skeptical humanism. "There is a certain strong and generous ignorance that concedes nothing to knowledge in honor and courage, an ignorance that requires no less knowledge to conceive it than does knowledge."[45] That impulse to recover "*Thaumas*" as the wellspring of knowledge situates Montaigne within the tradition running from Aristotle's catalyzing wonder to the je ne sais quoi of a later, more self-consciously developed aesthetic thought.[46] For Montaigne, accomplishing that legitimizing ignorance is a matter of thinking our way out of a preoccupation with causes. Men "leave aside the cases and amuse themselves treating the causes ... we know the foundations and causes of a thousand things that never were."[47]

For Montaigne, the case of Martin Guerre represents an especially consequential example of the risks of presumed epistemological certainty. Montaigne finds the imposture as the judge described it "so marvelous and so far surpassing our knowledge and his own, who was judge, that [he] found much rashness in the sentence that had condemned the man to be hanged. Let us accept some form of sentence which says 'The court understands nothing of the matter,' more freely and ingenuously than did the Areopagates, who, finding themselves hard pressed by a case that they could not unravel, ordered the parties to come back in a hundred years"—a temporizing solution apt in relation to a character and a case with such posthumous vitality.[48]

Given the essay's title, it's notable that Montaigne makes no mention of the wooden leg. Then again, it's hard to know why exactly the essay has the title it does. The piece is about the infirmity of human reason, but explicit references to "lameness" recur largely as a subterranean image pattern—a literary matter. "Montaigne keeps coming back to images of legs: the gouty legs of the prince supposedly cured by the 'wondrous deeds' of a priest; the thin legs of the French and the thick legs of Germanicus, both explained by horseback riding; the deformed legs of the pleasure-giving lame woman," Davis observes (122). And yet, the sidelong, displacing reference may be very much to the epistemological point. A knowledge founded on a claim to compass its own limits is a knowledge that will systematically exceed itself, evoking what it would compass always at its fringes. Legs according to such a performative epistemology are foundational just insofar as they stagger.

But Montaigne's most overt example of *les boyteaux* shifts beyond even that version of performative cognition insofar as it refers beyond speculative cognition altogether. "Apropos or malapropos, no matter, they say in Italy as a common proverb that he does not know Venus in her perfect sweetness who has not lain with a cripple."[49] Having evoked a full array of contradictory received accounts as to why that should be so, Montaigne concludes with the "woman weavers" whom the Greeks "decried . . . as being hotter than other women because of the sedentary trade they perform, without much bodily exercise": "What can we not reason about at this rate? Of these last I might also say that the joggling that their work gives them as they are thus seated arouses and solicits them, as the shaking and trembling of their coaches does the ladies."[50] The reflexive turn by which Montaigne claims knowledge, or, more formidably, the knowledge that he does not know, is here replaced by autoerotic excitation, his scrupulous avoidance of causal explanation now ecstatically realized by a solicitation whose source in this scene of technically mediated self-sufficiency remains as unknowable as God, the opaque origin of all causes. It isn't hard to feel the resonance between these weavers and the essay's own equally reiterative weaving of references untethered from semantic claims.

Is Montaigne's account of the Greek weavers a story about erotic desire enlisted in the purposes of servitude? Pleasure appropriating the means of production? It's hard to know, just as it's hard to know whether techne here is part of the self or alien to it. In that regard, the women weavers can be taken to figure the grounds of the social as a causeless cause that admits into the core of the formative social instance a reiterative otherness and finitude whose world-creative potential is inseparable from its power to resist from the outset any form of accounting or aggregation, an auto- and heterogeneous solicitation that is at once excitation and demand, invitation and compulsion. Rather than a "flat ontology," or an expansive web of correspondences, or even perhaps a coherently reflexive autopoietic mechanism, the women suggest the world as a weave of originary coincidence, one whose constellating power is inseparable from its disjunctive force, whose elements coincide with one another just insofar as they insistently fail to coincide with themselves: apropos *and* malapropos.

3

"No cause, no cause," Cordelia exclaims toward the close. No Shakespeare play enacts its own sourceless autopoietic condition as explicitly as the tragedy that begins with an ending, the king's staging of his own demise, and thus, according to the logic of autopoietic formation, no play so directly evokes the elusive conditions of world creation. If the tale of Martin Guerre figures the disruptive and consolidating intrusion of aesthetics within a worldly historical event, *King Lear* seems to exceed its own frame as a circumscribed aesthetic form, undoing itself as a mimetic representation in the process.

A subterranean current in every aesthetic formation, that boundary-dissolving effect has particular prominence in *King Lear* because of a convergence between form and content distinctive to the play, a convergence between art and sovereignty as equivalently self-founding categories. A thanatotic "crawl toward death," Lear's fantasy of overseeing his own demise equally enacts the self-compassment that institutes modern, postfeudal sovereignty as an autonomous formation, sovereignty as, in Jean-Luc Nancy's formulation, "an infinite institution that nevertheless includes within it the imperious necessity of the finite moment of its institution."[51] "I would unstate myself to be in due resolution," Gloucester says: Sovereign self-certainty demands just such a self-undoing.[52] In that regard, Lear's "mad" dissolution of his estate is the act that sets in train all those self-consuming, self-generating mechanisms the drama performs, explicitly autopoietic acts in which resolution and dissolution, cure and punishment, become indistinguishable. "I will punish home," Lear says, preparing to revenge himself against his own flesh and blood, healing and renewing the divisions that haunt him (3.4.16). "Is it not as this mouth should tear this hand / For lifting food to't" (1.1.117–19).

The illegibility of the "darker purpose" that propels the play's inaugurating scene, the fact that we can't know whether for Lear the scene is staged or real, whether the act is his own or a requirement of the structure that constitutes him, reflects the undoing of dramatic mimesis that necessarily attends that realization of drama as a fully autopoietic condition. With that, the play reveals the deeper grounds of the redounding, world-dissolving violence that afflicts it even as it opens the possibility of rethinking world according to conceptions of justice and rapport beyond the logic of equivalence that underlies that retributive violence.

The passage beyond limits that defines the play formally marks the drama at every level, from the most intimate constitution of sense experience to the play's conceptualizations of worldly equity and justice. Tragic wonder, what is prompted by the autopoietic return of what was cast out, amounts in *Lear*, as in *Guerre*, to a scene of recognition as missed recognition, correspondence as noncorrespondence, bearing on the very condition of affective being. As in the case of the prodigal husband, both the logic and the limits of autopoiesis as a founding structure are articulated in relation to that moment of formative incommensurability, a moment again evoked in terms of human disability. The divergent strands of the divided, double-plotted play converge at the point toward the close where mad Lear and blind Gloucester cross paths:[53]

Gloucester: O ruined piece of nature, this great world
Shall so wear out to naught. Dost thou know me?
Lear: I remember thine eyes well enough. Dost thou squiny
at me?
No, do thy worst, blind Cupid, I'll not love.
Read thou this challenge, mark but the penning of it.
Gloucester: Were all thy letters suns, I could not see one.
Edgar [*aside*]: I would not take this from report: it is,
And my heart breaks at it.
Lear: Read.
Gloucester: What? With the case of my eyes?
Lear: Oh ho, are you there with me? No eyes in your
head, nor no money in your purse? Your eyes are
in a heavy case, your purse in a light, yet you see
how this world goes.

(4.6.130–44)

Recognition occurs at the point of loss—"I remember thine eyes"—for through loss the possibility of substitutive exchange opens: "If thou wilt weep my fortunes, take my eyes" (4.6.172). But the scene moves beyond such a logic of specular exchangeability, beyond the play's concern with the reversibility of sight and blindness, interiority and exteriority. In Gloucester's extraordinary image of letters as suns the distinction between phenomenal and nonphenomenal dissolves altogether: sheer material inscription becomes the source of sensible, including erotic, existence: "Dost thou squiny at me ... blind Cupid," Lear says leeringly.

Phenomenality as inscription is a matter of seeing not with the eye, whether sighted or blind, but with the "case" of the eye, the eye's material housing as well as the prior condition for the possibility of visibility—to be in a "heavy case" is to be in a dire condition. The preoccupation with context and conditionality is related to the play's formal self-consciousness at the point where its plotlines cross, a self-awareness heightened by Edgar's reference to the truth status of the scene. Seeing as inscription takes place where a character sees and speaks as if from his or her own condition as a textual phenomenon, an impossible, boundary-crossing exteriority—an exorbitant framing—that amounts here to the very condition for the most intimate grounds of sensible being and rapport; "it is," Edgar declares. The scene in that sense translates autopoiesis from questions of the observer from the outset observed—a cognitive matter—to the unstable conditions of sense experience as such.

That passage via inscription to the determining limits of relation coincides with the play's most direct engagement with the world as a network of equivalencies. The encounter opens into Lear's evocation of social equivalence in the lines most often cited to support claims for the play's incipient democratic promise.[54] "A man may see how this world goes with no eyes. Look with thine ears. See how yon justice rails upon yon simple thief. Hark in thine ear: change places and handy-dandy, which is the justice, which is the thief?" (4.6.146–50). But if the promise of justice seems to lie here, the passage also leads seamlessly into Lear's fullest access of brute force: "I'll put it in proof / And when I have stolen upon these son-in-laws, / Then kill, kill, kill, kill, kill, kill" (4.6.181–83).

The turn to violence isn't a consequence of the reassertion of hierarchical force in the face of the prospect of demotic equivalency. Rather, it derives from the very logic of equivalence—its "proof." Justice as reversibility is a function of the world conceived as system of potentially limitless exchanges. Equity is, in that sense, an economic category, specularity bound to speculation: thus the allusion to money and coining. But because such a system of equivalencies is by definition infinite, it is inevitably staked on the violent exclusion through which it would constitute itself as a system at all. That violent limit is, in turn, illimitable—"kill, kill, kill, kill, kill, kill"—because it reiterates what it would resolve: Lear becomes the equivalent of the sons-in-law in the recourse to violence, becomes what he defines himself against. In a model of social and environmental justice based purely on equity, the irenic and the agonistic themselves become equivalent, one the reverse image of the other. In that sense, the violence in *King Lear* anticipates the catastrophic "reign of equivalence" Nancy associates with technological modernity—the darker lineage of an autopoietics of world.

Such a turn arises because the account of justice as reversibility loses track of what opened its possibility. For that encounter entailed a passage beyond equivalence or correspondence, an originative failure of specularity. Any possibility of justice derives, ultimately, not from equivalence, not even self-equivalence, but from equivalency's prior grounds, which is also to say not from the extrinsically imposed conception of the world as a speculative totality but from the world's prior condition as phenomenal appearance at all. "But who comes here," Edgar had said at the point of Lear's entrance in the scene:

	The safer sense will ne'er accommodate His master thus.
Lear:	No, they cannot touch me for coining. I am the King himself.
Edgar:	O thou side-piercing sight!
Lear:	Nature's above art in that respect. There's your press- money.

(4.6.80–87)

The "king himself" relates to money not only as circulatory equivalent but at the point of its production. They cannot touch me for coining: They cannot censure me for producing coins, for, as king, I am the sole legitimate coiner, outside exchange because I am its source. But then, too, you cannot touch me—you can't "catch" me, Lear will say in a moment—for the fact that I am coined, the one whose legitimizing face appears on the coin. And in that productive untouchability, a matter of being inside and outside exchange, lies the king's true touch, the test of truth and the grounds of all substitutions: To "touch" is to test gold by rubbing it on a touchstone.

The king's inside/outside, autonomic/heteronomic place is ultimately figured in a nontouching that is the grounds of touch and of sensibility. Lear's opening lines may or may not be an uncannily instantaneous response to Edgar's aside. "Who comes here? / The safer sense will ne'er accommodate / His master thus." The Arden note gives this as: "No one in his right mind would be got up like this." But the line needs to be read in more ontologically ambitious terms: Sense would not dress its own mastery in this way; sense—reason, but also phenomenal sentience—would not dress itself thus. The play's rich thematics of accommodation—accommodation in relation to clothing, disguise, to housing and "unhoused"ness, to the "thing itself" as "unaccommodated man"—condense in the figure of the sensible, of nature itself, as what is dangerously unhoused insofar as, failing to coincide with itself, it is nothing more than the fact of its own exposure. Like the limping punishment that arrives by not arriving, the tragedy figures being itself as sheer, mediatory relation, which is to say, as that which realizes itself by departing and imparting itself.

Such a version of recognition as nonrecognition reappears famously at the close, where it explicitly coincides with a crossing of the limits of the play itself.

> Thou'lt come no more,
> Never, never, never, never, never!
> Pray you, undo this button. Thank you, sir.
> Do you see this? Look on her, look, her lips,
> Look there, look there! [Dies]
> Edgar: He faints! My lord, my lord!
>
> (5.3.306–11)

What the king sees on Cordelia's mute lips is a corrective to his initial presumption: "Nothing will come of nothing: speak again." Nor are we in a position to say he sees nothing. The resources of cognizable meaning in the seemingly exhaustive opposition between reading literally—"Nothing will come of nothing, speak again"—and reading figuratively—Gloucester's nothing as mask or substitute for something: "come, let me see; if it be nothing, I will not need glasses"—between objective and subjective version of cognition, are undone here at the end in a performative crisscrossing of death's limit that dissolves the boundary between fictive and real, self and other, subject and object. The long awaited "recognition" occurs not in the play or in our mirroring cognizance of ourselves in the play but at the point where the play exceeds the audience's ability to compass its limits and thus to constitute itself in relation to that critical boundary. By virtue of that incognizable recognition, death reveals itself as that which is at once absolute—the unpassable limit of consciousness and being—and as what, insofar as it is the prior condition of cognition as such, has already been crossed, death itself proving at once determining of and subject to the missed encounter of autopoietic formation

"Are you with me?" Lear had earlier said. He is with Cordelia, and we are with him—never more proximate to him—insofar as he speaks from across an infinite divide; even as he crosses into our domain beyond the play's closure, we are drawn into his.[55] He doesn't return to us, nor does Cordelia return to him in the form of reverting conscience—tragedy shifts beyond such recuperative humanism here. Cordelia is with Lear as what was from the outset in his condition as sovereign subject at once inward and alien to him, which is to say, as definitionally phantasmatic. In Cordelia's returning to die again, the close enacts a sacrifice of consolidating, sacrificial exclusion, thus unbinding the logic through which the world is constituted as an object, whether the object of our violence or of our concern.

Because that enactment involves an actual crossing over of textual limits, the scene now implies a heteronomic self-relation that moves beyond the autopoietic redounds of self-protective force—the self-punishing claims of sovereign self-encompassment—that had propelled the drama. "Lend me a looking glass; / if that her breath will mist or stain the stone, / Why then she lives" (5.3.259–61). The crossing between animate and inanimate, human and inhuman, implied by the ghostly breath on the glass is picked up in the form of an address that takes in all including the audience—"Oh, you are men of stones!" (255)—and, in the Quarto, affects the most intimate resources of breath, materially inscribing the pure vocative of speech even as it opens speaking to its generative iterations beyond all characterological limits. "O, o, o, o," Lear declaims. "O, let him pass," says Kent. "O he is gone," says Edgar (5.3.309, 312, 314): We recall the Fool's reference to women making "mouths in a glass" (3.2.35). In the arresting emptiness of the mirroring stone our own response or nonresponse is also already inscribed. Neither subjective nor objective, belonging neither to one subject nor the other, shifting beyond the register of cognitive witnessing that characterizes systems theory, the inapprehensible breath in the mirror is what opens any such differentiation and thus amounts to something like subjectivity's causeless cause.

The ethics of the play is bound up with that causelessness—Cordelia's "no cause"—for it is the unbinding of causation that breaks the recursive mechanism that has propelled the tragedy's punitive and self-punitive course. Such a possibility, that which translates autopoietic recursiveness from the redounding violence of mastery and a limitlessly punitive conception of law to the promise of another relation to the world, depends on the ambiguous role of techne in the workings of autopoietic creation, understanding techne not as an object or thing but as the material trace within the structure of recognition—the breath in the mirror—that, eluding the human/nature distinction, resists speculative absorption. Bypassing apprehension insofar as it constitutes its condition, techne harbors the potential to unbind the spiraling violence of speculative equivalence even as it prompts the defensive reaction that threatens to perpetuate it.[56]

The self-consuming universe Lear evokes, where cure begets what it would resolve, can be seen to anticipate the proliferative disasters of our own ecological moment insofar as both participate in what Jean-Luc Nancy calls "the period of equivalences."[57] The play's engagement with coinage and exchange, especially in their relation to spiraling violence, predicts the extension of capitalism's general equivalence to the point not just of an exploitative alienation of the world, as Marx had it, but of a radical "indifference to the existence of the world and all its being"—an apt enough description of the tragedy's bleakest evocations.[58] Conceiving the world as an ecological "web of reciprocal causes" will not take us beyond that reign of equivalency.[59] Acknowledging "the existence of the world and all its being" instead means thinking the world outside speculative equivalency and the totalization such a conception implies, thinking stone, feather, art, as the causeless cause of existence as constitutive incommensurability.

A habitus understood in such terms is not defined by a given space, whether social, natural, or cosmic. For to conceive relation as a function of a given, especially an inclusively given, world is already to abolish relation and thus to abolish world, in the transcendental posture that claims to make that space knowable—the sovereign posture of self-seeing that sets all the play's woes in train. What occurs at *Lear*'s close doesn't *take* place, that is, its character as event doesn't transpire in a given subjective or objective space. Rather, it opens space—creates and decreates world—in the finite and self-divided interval that constitutes the possibility of any rapport.[60] Such an opening or habitus suggests an ecology—a logos of *oikos* or home—that conceives justice and relation from the vantage point of the world's primordially "unhoused" condition, which is to say from no "vantage point" at all.[61] That means reimagining finiteness not as what we must tragically face and thus can sovereignly presume to face but as what beyond all speculative investment already forms us in the present condition of exposure.

2

Timon of Athens, Autopoiesis, and the Political-Aesthetic Foundations of the Polis

It is no surprise that a play that describes a world where all that's solid melts into air, in which every human bond proves limitlessly fungible, should at an important junction feel the need to evoke the defining bounds of the social domain. Marking his turn mid-drama from *philanthropos* to *misanthropos*, Timon signals the play's movement beyond the Athenian polis with an apostrophe to the city wall:

> *Enter Timon*
> Timon: Let me look back upon thee. O thou wall
> That girdles in those wolves, dive in the earth,
> And fence not Athens! Matrons, turn incontinent!
> Obedience fail in children! Slaves and fools,
> Pluck the grave wrinkled senate from the bench
> And minister in their steads![1]

That such a boundary is described after it has been passed and as a call for its obliteration hardly rules out its potentially foundational status. "Not the act of tracing boundaries, but their cancellation or negation is the constitutive act of the city," Giorgio Agamben writes.[2] Precisely because it is constitutive, the boundary of the socius will only ever be knowable at the point of having passed it, as a limit and no limit. In that sense, echoing the peculiar formal character of a play that continues on well beyond the demise of its hero, such a midcourse passage beyond the limit is consistent with the movement of self-surpassing by which the modern socius, untethered from secure transcendental reference, forms itself from within itself, in and against itself, by virtue of what Claude Lefort terms the "enigma" of its "internal/external articulation";

consider, for instance, the way the threat of brute, wolfish "nature"—a retroactive construction of civility—functions for Hobbes as the commonwealth's founding cause.[3] If *King Lear* seems to gather the world beyond into its own negational autopoietic forming, ultimately suggesting the inhuman conditions of sense experience, *Timon of Athens*, with its focus on the founding dimensions of the friend-enemy distinction, brings its autopoietic form to bear on the founding of the polis.[4]

For Agamben, that logic of the included exception defines Western sovereignty as a specifically biopolitical category. Sovereign authority constitutes itself by way of its symmetrical counterterm "bare life," an abjected and phenomenally ambiguous creatural domain that will remain forever at once within and beyond the space of law. Timon's reference to citizens as wolves chimes with Agamben's account of the human/inhuman *loup-garou* as an exemplary instance of such a threshold condition.[5] Still, it's not clear how a biopolitical account, even one that extends to the limits of human abjection, is adequate to a drama whose reiterative endings reach beyond the domain of the living to a sea-lapped grave that weeps with no witness by and an epitaphic inscription that "interprets for" those incapable of interpreting it for themselves.

That autonomic dimension is related to the play's preoccupation with the speculative reach of economy, the uncannily agentless agency of money. But the play's various exorbitancies, its status as a tragedy that continues on beyond the titular hero's death just as it has continued on in the heated commentaries it has spawned, is equally and more fundamentally related to its character as an autonomous, self-founding, and self-exceeding aesthetic formation: to the *poetics* of autopoiesis.

Why should a play focused on the condition and limits of social rapport, limits evoked via extremities of human degradation, also be preoccupied with the nature and status of art, the drama framed as it is by a paragonal debate between poet and painter and returning throughout its course to questions of "taste"? For all its apparent association with cultural refinement, that political-aesthetic dimension runs deeper, I will argue, than the exclusionary logic of bare life or the speculative logic of economic relation, for it engages the necessarily prior means through which relationality as such does and does not assume phenomenal form. We should note what is most notable about the scene on the outskirts of Athens: the fact that it takes the form of a rhetorical apostrophe, and one directed neither to a subject nor to an object, neither to what falls within nor beyond the limits of the polis, but to the embodiment of division itself. "Oh thou wall," Timon declaims, addressing his gaze, we might imagine, to the theater's own defining partition. Bearing on the inscriptive force of history as well as the force of our own interpretive engagement, the play's reflexive inscription of its own material conditions locates the site of political formation beyond either Agamben's narratives of sovereign origination or the stable referential opposition that structures Luhmann's account of autopoietic systems, in relation to the incommensurable grounds of reference as such. In doing so, the play intimates a conception of socius and world more fundamental than the play's governing distinction between autonomy and communal bonds.

1

To find our way to the gates of Athens and the larger questions of political foundation that scene engages, we first need to tease out the relations among aesthetics, economy, and the logic of affective bonds as the play explores them. Though there is scholarly consensus about the play's concern with such bonds, what that means remains a notably open question. Although criticism has traditionally dwelled on the starkness of Timon's transformation from philanthropic to misanthropic extremes, more recent accounts have made claims for continuity. For Coppelia Kahn, Timon's apparent magnanimity involved a covert social power all along, an aggressive potlatch-like expenditure serving to tighten his hold over others via a debt that can come to feel inescapable.[6] "I'll pay the debt and free him," Timon says of the companion locked away in debtor's prison. "Your lordship ever binds him," replies the messenger (1.105–6). On the other hand, Ken Jackson has read Timon's largesse and refusal of reciprocity—"there's none / Can truly say he gives if he receives"—as a religious impulse aimed at outstripping the economy of exchange altogether, implying a commonality between the misanthrope's world-annihilating desires and the exorbitance of the pure, impossible gift, the gift that refuses the mutuality that underlies normative ethics (2.10–11).[7] Such uncertainties about the trajectory of the hero's life reflect opposed conceptions of the aim of social desire as such in the play, construing it on the one hand as an impulse toward wholeness, whether in the form of a socially consolidating philanthropic bond or a narcissistically consolidating return on one's social investments, and, on the other, as a drive toward division and dispersal either in the form of misanthropic animus or the self-exceeding expenditure of the pure gift.

Those ambiguities arise because, in a world in which identity as such is conceived as a function of exchange, the very distinction between expenditure and return, division and wholeness, dissolves. Jackson is right to read the play as problematizing exchange. But casting that ambition as Timon's desire to "seek the impossible gift outside exchange"—a matter of "pushing toward the gods in more authentic fear and trembling"—may give insufficient weight to one of the drama's more subversive ironies: that such incommensurability informs the "circular economy of exchange" itself.[8] The aporias of exchange become apparent precisely at the point where a return is demanded and all debts are due. After Timon has been shown to have vastly overspent himself in his munificence, his various lenders, themselves the beneficiaries of his largesse, call in their loans, a situation that prompts the following exchange between the lords' servants:

> Titus: I'll show you how t'observe a strange event.
> Your lord sends now for money?
> Hortensius: Most true, he does.
> Titus: And he wears jewels now of Timon's gift,
> For which I wait for money.
> Hortensius: It is against my heart.
> Lucius' Servant: Mark how strange it shows.
> Timon in this should pay more than he owes,
> And e'en as if your lord should wear rich jewels
> And send for money of 'em.
>
> (8.18–27)

It is as if, according to the "strange event" of such an economy, one owes for what one has already given or seeks payment for what one already possesses.

The erosion of what constitutes one's own at such a moment isn't simply a function of reaching a point of speculative overextension. It arises because identity as such is constituted within those structures of exchange. Defined in irreducibly relational terms, identity becomes an effect of originative indebtedness, expenditure the exorbitant leap that opens the play of relational differences within which any conception of identity or possession becomes possible. When Timon's creditor feels compelled to "call upon his own"—to demand his return—we need to read the demand simultaneously as the end point of exchange and its origin, the *meum* of "his own" constituted at the point of such a return; in that sense, one does indeed owe for what one already possesses. Do we read Timon as seeking self-aggrandizing returns on his expenditures or as genuinely self-abrogating? The problem is that the two can't be separated. Timon "outgoes / The very heart of kindness" because kindness—compassion but also equivalency—is constituted in the movement by which exchange initiates itself in excess of itself (1.277–78).

The temporal disturbance implied by such an indistinction between expenditure and return, beginnings and ends, can be felt throughout the play and defines the social bond in its most intimate, affective dimensions. Luke

Wilson describes the way the confusion of sequence and agency in the passage in which the servants speculate on Timon's speculations "swallow[s] up the time" that credit requires for its "strategies of delay and precipitancy."[9] Such untimeliness defines affective experience as such in the play. "O, what a precious comfort 'tis to have so many like brothers commanding one another's fortunes! O, joy's e'en made away ere't can be born: mine eyes cannot hold out water, methinks. To forget their faults, I'll drink to you," Timon declares (2.99–102). Anticipating the tragedy's larger transformations, the odd turn from ecstatic brotherhood to a joy lost before it has begun occurs because such abortive precipitance is the defining condition of a bond that begins at the point of its depleting expenditure, a bond that underlies all those forms of affective attachment that are only strengthened through their breach.

Reflecting the effects of early capitalism's reduction of all human bonds to structures of exchange, such exorbitance equally defines affective attachment in its most ineffably humanistic forms during the era. In "Of Friendship," Montaigne articulates a pure union unavailable to any measure beyond itself: "Our friendship has no other model than itself, can be compared only with itself"; friendship is, by such an account, the undefinable definition of ipseity: "If you press me to tell why I loved him . . . : Because it was he, because it was."[10] As it is for Timon, such a singular union is absolute insofar as it is a function of the expenditure that makes it what it is. It is, Montaigne writes, "the one who receives the benefit who provides the benefit"; when impoverished Diogenes is finally compelled to ask for money from his friends, Montaigne tells us, "He used to say that he asked it back of his friends, not that he asked for it."[11] Such "precipitancy," as Montaigne puts it—from the outset a returning—marks the attachment from its beginning. They knew each other before they met, Montaigne insists:

> Beyond all my understanding, beyond what I can say about this in particular, there was I know not what inexplicable and fateful force that was the mediator of this union. We sought each other before we met because of the reports we heard of each other, which had more effect on our affection than such reports would reasonably have: I think it was by some ordinance from heaven. We embraced each other by our names.[12]

At once self-identical and mediated, singular and self-divided, the true bond is measured by its capacity to appear momentously in advance of itself.

"We embraced each other by our names"; "The world is a word," Flavius declares (2.2.147). The groundless character of affective experience thus conceived brings us back to the question of the play's reflections on its own literariness. Laura Kolb persuasively connects the foregrounding of art in the play to the credit economy the play articulates, an untethered system in which value becomes a function of persuasion.[13] But beyond such rhetorically instrumental ends, it is through art's creationist dimensions—its power to constitute worlds—that the play thinks through the riddle of contingent origination, the formative power of what, like money and art, should be merely supplementary or substitutive.

Timon opens with a poet and painter offering their metadramatic observations on what we are about to observe. "Or poesy is as a gum which oozes / From whence 'tis nourished. The fire i'th' flint / Shows not till it be struck; our gentle flame / Provokes itself, and like the current flies / Each bound it chafes," the poet declares at the opening (1.21–25). Poetry maintains its self-generative power by virtue of its own chafing impediment. Provoking itself in the movement by which it exceeds its own bounds, poetry is self-sufficient by virtue of its self-dividedness. The contradiction crystallizes in a single image:

> My free drift
> Halts not particularly, but moves itself
> In a wide sea of wax—no levelled malice
> Infects one comma in the course I hold—
> But flies an eagle flight, bold and forth on,
> Leaving no tract behind.
>
> (1.45–50)

Precisely in its claim to self-sufficiency, such free poiesis can't help but evoke what it sets itself against. To wax is to boundlessly increase, but it also brings to mind the resistant ground upon which the poetry must inscribe itself, ensuring that it does in fact "leave a tract behind": The passage anticipates the wax on which Timon's epitaph—the source of his story—will be imprinted. Even the comma, purportedly untainted by extrinsic aims, evokes its own caesura-like impediment; it is "almost as though the comma were the infection, potentially disrupting the 'flight, bold and forth on,'" Jowett's editorial note on the line observes.

The indistinction between poiesis and its impediment also means an indistinction between the play and the chafing response it provokes. The drama can feel like nothing more than an ongoing commentary on itself, Lesley Brill points out: "Nowhere else in Shakespeare's work do so many characters comment so copiously upon such scanty action"[14]—thus the consistently allegorical feel of the work. The proliferation of critical debates around the play reflects, says Brill, the play's own multiplication of divergent views around its fatally imperfect world; the play is, in that sense, the problem of its interpretation. At the same time, because the work *is* that self-commentary, because those chafing supplements are constitutive, in *Timon* the world itself is a matter of aesthetic taste, a place where brooks are "candied," youths pursue their "sugared game," and Timon declares he has "the world as [his] confectionary" (14.227, 260, 261).

Aesthetics' indissociable relation to sense experience is realized in the promiscuous expansion of the play's references to "taste." The play's initial Chapmanesque "banquet of sense" is echoed later in Timon's mock feast of stones and smoking water. Aesthetic sensibility escalates to eating, eating escalates to cannibalism, with reiterated allusion to Timon's debtors consuming him: "O you gods, what a number of men eats Timon, and he sees 'em not," Apemantus declares (2.40). Framed initially in relation to art, such moments inevitably reflect the audience's own lurid investments in the tale of ruination. "I feed / Most hungrily on your sight," Alcibiades says to Timon, as if with a glance toward our own consuming gaze (1.256–57).

To "feed" on "your sight" is to feed on the sight of you and on your viewing of me, a reflexive ambiguity translating Luhmann's account of the originative status of the observer observed into especially graphic terms, bringing to mind as it does the play's preoccupation with cannibalism—a matter of eating one's own.[15] With the indistinction between sense experience and one's resistant response to it, interpretive judgment is always at risk of reverting upon itself, a prospect represented in the play by the ambiguities of the act of apprehension:

Timon:	Wilt dine with me, Apemantus?
Apemantus:	No, I eat not lords.
Timon:	An thou shouldst, thou'dst anger ladies.
Apemantus:	O, they eat lords. So they come by great bellies.
Timon:	That's a lascivious apprehension.
Apemantus:	So thou apprehend'st, take it for thy labour.

(1.205–12)

Literal and figurative consumption—consuming the lords' wealth as consuming the lords themselves—is here allied as well with sexual devouring—what makes the ladies' bellies great—with all these acceptations crystallizing in the play on "apprehension," a term that refers at once to the literal sense of taking hold and to the cognitive sense of grasp. The consuming joke reverts against Timon insofar as his reading repeats the sexual apprehension it would grasp, his labor yielding nothing but its own laboring.

The slippage between physical and interpretive grasp condenses in "take," a word that recurs throughout the play and that conveys the redounding character of apprehension in its mingling of transitive and intransitive uses: The soldier at the close will actively and passively "take" Timon's epitaphic "character" with wax. After Timon praises the masque, the "First Lady" of the performers responds, "My lord, you take us even at the best," to which Apemantus retorts: "Faith; for the worst is filthy, and would not hold taking," a phrase that leaves it uncertain whether they are taking or being taken, grasping or being grasped (2.148–50). The point isn't simply that interpretation is analogous to literal grasping; it's that they are ultimately indistinguishable, aesthetic apprehension enacting the condition of phenomenal apprehension in its own chafingly self-divisive movement, an enactment that leaves one literally and figuratively eating nothing: a banquet of smoke and stones.

That convergence between art and phenomenal experience lies behind the implicative force of the drama, the way its divisions—agonistic divisions between characters, as well as the play's own bifurcated form—reproduce themselves in the notable critical divisions the play has spawned. Precisely because the world *is* a word, one is never in a position to comprehend it as *merely* a word, to assume the extrinsic or transcendent position from which one would confidently know fiction as mere fiction. Timon recognizes flattery for what it is—"well mocked" he says to one of his suitors, just as he praises poetry for its being true to the falseness of the world—even as he becomes absorbed by its lures (1.177, 163). You have "entertained me with mine own device," Timon says to the masquers, caught up as he has been in a theater he himself has contrived; it's a line in which the consuming and consumed audience might recognize itself (2.146). As a condition of aesthetic consciousness, one remains ever knowing and absorbed, taken by what one presumes to take. In that sense, the play locates us squarely within a mocking political-aesthetic space not unlike our own political moment, where every provocation is a lure and every act of resistance seems condemned to renew the contagious drama it seeks to oppose.

2

Thus far, I have been speaking of the analogy between the play's economic and aesthetic concerns, domains comparable in their autonomous and self-referential workings. But art is its own sphere and has its own ideological function. Precisely in the self-consciousness with which it reflects on its autopoietic formation—precisely in its status as aesthetic *consciousness*—aesthetics, for all its evident exorbitancy, also serves a consolidating function, a function that lies at the heart of its political vocation. To see that, we need to return to the outskirts of Athens. An outright call for the destruction of every bound, the scene staged at the limits of the polis amounts at the same time to a delineating caesura—a "comma"—at the center of a drama in which autonomy and heteronomy, inside and outside, wholeness and difference threaten to dissolve into limitless play. Through that internalized limit, the play aligns Timon's act—a thematic issue—with its own condition as an aesthetic form, the self-dissolvative and self-generating movement by which the work reflexively exceeds itself in order to define itself. In that regard, the tragedy's formal oddity, the fact that it continues on beyond the titular hero's death, is of a piece with its character as an autonomous aesthetic form—indeed, one might argue that in its peculiar, self-exceeding status, the tragedy signals the historical emergence of the notion of aesthetic consciousness.

Insofar as aesthetics thus claims the capacity to make autopoiesis available to consciousness, to translate it into a speculative form, it also makes aesthetics available as a figure for foundations generally, including political foundations. Precisely in its manifest groundlessness, or its capacity to make groundlessness manifest, aesthetics assumes an ideological function, giving apprehensible form to the founding division Carl Schmitt terms "sovereign nomos," the constitutive ban, like the "multiplying bans" Timon levels against the citizens, through which the state defines itself.[16] Translated into a speculative and thematic condition, aesthetics becomes equatable: Precisely as autonomous, art models a world beyond itself. At the same time, the fragility of that speculative operation, the inevitability with which autopoiesis exceeds and troubles the psychic and political totalities it sustains, opens the prospect of a disturbance of political foundations more thoroughgoing than any thematic threat the drama might evoke.

To bear out such a political-aesthetic claim, it's helpful to place it in relation to Agamben's arguments for the biopolitical foundations of the polis. Though Agamben speaks of bare life—the liminal life that can be killed but not sacrificed—as sovereignty's ultimate point of reference, insofar as sovereignty as much as bare life is constituted through that founding division—insofar as both are an effect of it—it would be more accurate to say that it is difference as such—the wall—to which sovereign power must lay claim. It's not the onslaught of beasts that Timon evokes but a peculiarly generative dissolution of every bound and bond:

Piety and fear,
Religion to the gods, peace, justice, truth,
Domestic awe, night rest, and neighborhood,
Instructions, manners, mysteries, and trades,
Degrees, observances, customs, and laws,

Decline to your confounding contraries,
And yet confusion live!

(12.15–21)

Confusion lives on beyond the end because it is the source of what it undoes. "Confound"—a term that recurs in the play like a guiding thread—crystallizes the logic of a formative dissolution of distinctions. The determining force of such malice is apparent in the reversibility of the oppositions it creates. "Henceforth hated be / Of Timon man and all humanity!" Timon declares, at once subject and object of the hatred he calls down on all (11.103–4); "Here lie I, Timon, who alive all living men did hate," reads his epitaph (17.73). The self-referentiality of Timon's animus, the fact that he includes himself in his denunciations and hatefully prays that his own hatred might continue, betokens a malignancy that exceeds every agency, including the misanthrope's own, if malignancy remains the term for that form of prevenient causation (14.22, 12.39–40).

To make such world-creative claims for "confounding" poses an obvious epistemological question. If the inside/outside, friend/enemy distinction is the condition of political identity and indeed of identity as such, how is that division cognizable at all? The problem is implicit in a contradiction at the core of Agamben's origin story. Because the mechanism of inclusive exclusion is constitutive, it is itself without "positive content," without referent because it constitutes "the structure of reference in general," Agamben asserts.[17] And yet, in the course of Agamben's various narratives, that founding mechanism nevertheless does assume an object: "bare life" as the correlate, indeed, "the first and immediate reference," of the sovereign ban.[18]

Bare life is without essence: It is a definitionally indeterminate category, existing always in the "zone of indistinction" between *zoe* and *bios*, natural and politically determined life. And yet by granting phenomenally substantialized form to that indeterminate condition—by translating it into sovereignty's object at all—Agamben is able to convert a repetitive and necessarily open mechanism—the process by which, as in *King Lear*, sovereignty at once constitutes and endlessly consumes itself, divides itself to make itself whole—into a drama of "capture" in which life is caught within law and through which the sovereign becomes at once the effect of the ban and its arbiter.[19] Recuperating a purely formal operation in these agonistic terms, Agamben is able to recast the entire Enlightenment project and its preoccupation with thinking the incognizable limits of cognition—what is "in force without significance"—as one more turn in the dire advance of biopower's expansive inscription of life, Kant becoming a way-station in the inevitable biopolitical trajectory toward the Nazi death camps.[20] Moreover, once sovereignty's nonground becomes "capturable" as a phenomenally legible form, a reiterative operation becomes capable of narrativization. What had bypassed the claims of reference in its unbounded operation now becomes fully determinate spatially and temporally, traceable in an epochal arc culminating in the delimiting confines of the death camp.

What drops out of Agamben's account is how such an indeterminate condition becomes phenomenally available at all. Ultimately, that is a political-aesthetic rather than a biopolitical matter, and it is only by engaging the aporias of that rhetorical process of phenomenalization that we can loosen the forbidding hold of Agamben's sweeping narrative of political capture. To open that political-aesthetic dimension in all its ambiguity, we need to read the scene at the gates in relation to the formative space of theater itself. The moment Timon redefines himself by passing beyond the limits of the polis is also a moment of self-recognition for the theater audience. Of Timon's reference to the Athenians as wolves, Jowett writes: "The line would have offered the actor in the Jacobean public theatre an opportunity to invoke the galleries of spectators within the theatre walls, which would have extended from the tiring-house wall. This would make a theatrical point of the paradox: the audience is at once with Timon outside the walls of Athens and enclosed by the theatre walls, perhaps implicated as 'wolves.'" Indeed, a scene later, Timon will direct his curses against the citizens "both within and out that wall" (12.38).

An ethical matter, such a divided condition is at the same time the cognitive condition of the audience's recalling itself in such a self-consciously metatheatrical moment. The mechanism by which the play reflexively constitutes itself as aesthetic form repeats the condition of consciousness, consciousness understood as the division and loss entailed in the very movement by which it reflexively comes to.[21] Through that performative enactment—consciousness as the consciousness of one's reiterated blindness and absorption—difference is translated into the subject's groundless ground, cognized as what remains incognizable.

By construing aestheticization in these foundational terms, I'm embracing an account quite distinct from Agamben's own historicizing account of the category. For Agamben, modern and especially post-Kantian aesthetic consciousness amounts to a fallen condition marked at once by empty, self-negating self-consciousness and a subjection to a biological "metaphysics of the will," the latter a particularly surprising conclusion given Kant's explicit separation of aesthetic experience from appetitive need or interest.[22] Agamben's counterintuitive account derives from his assumption that aesthetics is a function of the historically empirical division between creative poiesis on the one hand and spectatorship on the other, the former associated by virtue of such a separation with individual creative will, the latter with empty reflexivity.[23] With that fall into "the desert of *terra aesthetica*," modern consciousness in its dividedness loses touch with the nonbiological, nonsubjective "ek-stasis" by which Greek thought had sustained a more "authentic" unity between art and world.[24]

Even aside from the question of how much that vision of a once unified sense is itself a projective function of the divided condition he laments, we can note the ways in which Agamben's account fails to register the deeper stakes of Kantian aesthetics. For Kant, aesthetic judgment holds the potential to bridge the apparently irreconcilable domains of his analytic of the human faculties precisely because its reflective form of apprehension bears on the foundations of phenomenal perception as such.[25] The reflexive movement of that judgment, a judgment that posits what it at the same time discovers, implies a division and other-directedness at the origins of sense perception that dissolves the distinction between creation and reception, denying even as it affirms Agamben's vision of an authentically unified and immediate sense.[26] With aesthetic judgment, Jacques Rancière writes, "*poesis* and *aesthesis* stand . . . in immediate relation to each other. But they relate to one another through the very gap of their ground. They can only be brought together by a human nature that is lost or that is yet to come."[27]

They relate through the "gap of their ground," and only insofar as they compass that gap *as* their would-be ground, and it's in relation to that requirement that the contradiction at the core of aesthetic consciousness becomes apparent. The necessity and impossibility of such a founding relation to relation in the constitution of sense experience is apparent in the uncanniness of Timon's apostrophe as a consciousness-forming instance. "The grand vocatives of Timon's soliloquies in this passage shape out the physical structure of the building as though the audience heard on behalf of the building, and yet as though human beings were absent," Jowett writes in his introduction to the play text (14). Constituted via the vicarity of the theatrical medium itself, consciousness forms itself as that which is by definition not there and cannot hear. As Paul de Man has demonstrated, that originative displacement is a function of the elliptically formative character of apostrophe. While through that rhetorical address the subject grants life—a capacity to hear and respond—to what is inanimate, it is at the same time only as a consequence of that arbitrary positing that the subject first knows itself as animate being, consciousness emerging after the fact as a function of an impossible crossing between animate and inanimate.[28] The missed encounter of tragic recognition becomes the condition of our own coming-to, a riven condition we can only repeat in the process of making it knowable—a matter of encountering the force of our positing as what will have preceded us.

The object of Timon's apostrophe—the wall—ultimately amounts to a figure for the constitutive division of the figurative address itself, a rhetorical effect that makes any such objectification possible. The address anchors the drama most securely to an immediate referential ground—the literal space of the theater—but only by virtue of the "gap" of that ground, as a groundlessly reiterative rhetorical effect. As the contradictory condition of sense certainty here indissociable from the mediating structure of theater itself, such an effect inscribes the auditor as much as the hero in its formative displacements. As *poiesis* converges with *aesthesis*, so character and audience converge, together suggesting the unstably breached/unbreached foundation of the polis.

3

Timon's apostrophe aligns the failure of political-aesthetic formation with a referential instability that troubles autopoiesis itself as a coherent operation. In the very of movement of its necessary self-reference—theater's grounding itself in relation to itself—apostrophic address exceeds itself and exceeds the system it would assure, undoing the distinction between auto- and heteroreference on which Luhmann's account of autopoietic formation is founded. Beyond either the knowing unknowingness implied by aesthetic comprehension or even the necessary blindness implied by one's inscription within an autopoietic structure—the correlative of a clarity of sight granted by a view from without—the disturbance entailed in Timon's address extends to our own capacity to grasp autopoiesis as a determinate structure, transforming the "gap" of tragedy's missed encounter into something more undecidable, and less assimilable, than the temporizing interval Luhmann sees as the necessary condition of reflexive formation. Drawing us into its reach, such referential instability implies a formative contingency and historicity more radical than the narratable sequence of autopoietic forms, even as, in thus troubling the distinction between auto- and heteroreference, it suggests the possibility of reconceiving sociality beyond the familiar oppositions between autonomy and heteronomy, freedom and subjection.

That sense of a hitch in the process of autopoiesis, something inassimilable to the resolvative/dissolvative mechanism of political-aesthetic formation, is intimated by a curious detail of the scene at the gates of Athens. Timon doesn't enjoin the wall to fall; he calls for it to "dive in the earth," a catachresis suggesting a conjunction between the aqueous and the earthly, between fluid exchangeability and solid ground.[29] The sense of something subterranean and forced will recur moments later in Timon's discovery of buried gold and then in the repeated allusions to Timon's own burial. Those reiterated allusions to inhumation, each as we will see describing a form of living burial, suggest something more untoward than the forms of constitutive negation that define the socius as a comprehensible system of exchange. In its later stretches, as the play extends itself beyond its hero's end and thus in a sense beyond its own logical end, it will be through and against that inassimilable catachretic remainder that the play will attempt to reconstitute a stable biopolitical narrative of political authority, one that would reclaim the sovereignty of human intention.

Soon after his call for the walls to drown in earth, digging for roots, Timon uncovers buried gold, a moment that suggests the relation between exchange value and unknowable grounds:

> Earth, yield me roots.
> *He digs*
> Who seeks for better of thee, sauce his palate
> With thy most operant poison.
> *He finds gold*
> What is here?
> Gold? Yellow, glittering, precious gold?
> No, gods, I am no idle votarist:
> Roots, you clear heavens. Thus much of this will make
> Black white, foul fair, wrong right,
> Base noble, old young, coward valiant.
> Ha, you gods! Why this, what, this, you gods?
>
> (14.23–31)

As if conjured by his allusion to nature's "operant poison," Timon discovers precisely what he seeks to abjure, the power to confound every distinction, now specifically associated with gold's transformative properties. It's a complex moment. Gold, that which sets in motion the movement of perverse substitutability, is itself caught in an instance of perverse substitution: roots for gold, the root of all evil. This can be read as a comment on the illimitability of exchange: everything inscribed to the root in speculation. The gold is, in that sense, previously buried currency: Timon himself will rebury the discovered gold again in a moment (13.10). Then again, the "yellow, glittering" gold could also be earth's own ore—Timon will speak of it as "damned earth" (14.42). "Yellow, glittering" gold recalls in its creative potential Timon's reference to the "blessed, breeding sun" at the start of the address.[30]

Marx, who was drawn to *Timon* as a critique of capital's "confounding of all human and natural qualities," provides a clue to that contradictory relation between creative source and substitutability:

> Just as all the qualitative differences between commodities are effaced in money, so money on its side, a radical leveler, effaces all distinctions. But money is itself a commodity, an external object, capable of becoming the private property of any individual. Thus social power becomes a private power in the hands of a private person. . . . Modern society which, when still in its infancy, pulled Pluto by the hair of his head out of the bowels of the earth, acclaims gold, its Holy Grail, as the glittering incarnation of its inmost vital principle.[31]

The inclusive exclusion Agamben associates with sovereignty is here realized as the basis of exchange in the form of a substanceless mediating term—what institutes commodification—that is nevertheless capable of itself becoming an objective commodity, money thus becoming what remains beyond the "great social retort" of circulation as its cause even while incorporating itself within that retort. The magic of that "sensible supersensible" form is conveyed for Marx as for Shakespeare as a reflective "glitter" that locates gold's mysterious power at once within and outside itself.[32] Instituting value as pure relationality, money becomes infinite, merely strengthened by the attempt to negate or pass beyond it; just as Timon discovers gold where he would turn from it, so according to Marx least "able to withstand" money's "alchemy" are those "sacrosanct things which are outside the commercial traffic of men."[33] And yet, by virtue of that alchemy, infinite exchangeability—intangible "social power" as such—becomes possessable, "a private power in the hands of a private person."

Like Agamben's claims for the sovereign's power to "capture" what produces him, the self-inclusive property of the money-form is a fiction. But in this instance, it's a fiction that intimates the limits of such fictionalizing narratives. Marx's description converts a boundless structural effect into a story of origination, but an opaque one. Did modernity "in its infancy" seek gold from the "bowels of the earth" because of its value? Or does value emerge with gold from the earth? Gold is a story of buried beginnings because that question remains unanswerable.[34] Figuring the limits of figuration, glittering gold represents that which can't be assimilated to the movements of narrative and economic substitution it sets in motion, a permanent hitch in the story of its own unfolding. Marx's allegory of origins involves its own substitutive confusion—Pluto, god of the underworld, for Plutus, god of wealth—and thus a buried source at its source. To constitute exchange, even infinite exchangeability, as a comprehensible phenomenon—to make it narratable at all—requires a prior suppression or burial, one that repeats the division it would allay and thus renews the strangely possessive hold of what one would cognitively possess. "Thou'rt quick; / But yet I'll bury thee," Timon declares as Alcibiades and his soldiers approach, reburying the uncannily living material even while holding back a small sum—"stay thou out for earnest" or a pledge—as the remainder that will once again set in motion the play of exchangeable signs (14.45–46, 48).

The equivocal living burial that institutes exchange value in the economic domain will be echoed in the ambiguities of Timon's own burial, a no less ambiguous event bearing on the play itself as a signifying system. The founding paradox of inclusive exclusion reaches a kind of sublime limit in the enigma of that ending: How does a hermit bury himself? Robert Darcy resolves the question by suggesting that the grave is actually an empty cenotaph, that Timon has staged the close, thus bearing out the play's own empty theatricality.[35] Given the ambiguity of limits in the play, it might be more accurate to say that Timon is and isn't buried in his grave. Can you bury yourself? In a sense he has done that all along. "Timon is dead, who hath outstretched his span," reads the epitaph (16.3); in fact, the backward-looking, self-exceeding misanthrope for whom joy is "made away er'it can be born" outstretched himself from the outset.

That paradox of constitutive overreaching crystallizes in the epitaph, a text that conjures the hero's voice beyond his death even as it draws us toward the generative origins of the play as a historical document. The sense of something interpretively irresolvable recurs in both of the two versions of that memorial offered in the play. The first account of the grave is provided by a soldier seeking Timon in the woods:

> By all description, this should be the place.
> Who's here? Speak, ho! No answer?
> *He discovers an inscribed tomb*
> What is this?
> 'Timon is dead, who hath outstretched his span.
> Some beast read this; there does not live a man.'
> Dead, sure, and this his grave. What's on this tomb
> I cannot read. The character I'll take with wax.
> Our captain hath in every figure skill,
> An aged interpreter, though young in days.
>
> (16.1–8)

The speech is problematic. If the soldier can't read the inscription, what were the words he just recited describing Timon's life? They might be the soldier's own comments, but most editors agree that, given their misanthropic tenor, as well as their couplet form, the lines probably are an epitaph. Editors have resolved the issue by reference to the apparently unfinished state of the play, suggesting that the spoken couplet would have been eliminated in revision. But as Jowett's note observes, "a superfluity of epitaphs seems oddly appropriate to the play and to the character," and here what is notable is the enigmatic correspondence between the two elements, the couplet's "some beast read this" and the soldier's illiteracy, as if he were, in fact, the message's correct destination.

The contradictions are just as stark in the second iteration. With Alcibiades at the gates and threatening the destruction of Athens, the soldier arrives with the news of Timon's death:

> Soldier: My noble general, Timon is dead,
> Entombed upon the very hem o'th' sea;
> And on his gravestone this insculpture, which
> With wax I brought away, whose soft impression
> Interprets for my poor ignorance.
> *Alcibiades reads the epitaph*
> Alcibiades:
> 'Here lies a wretched corpse, of wretched soul bereft,
> Seek not his name. A plague consume you wicked caitiffs left!
> Here lie I, Timon, who alive all living men did hate.
> Pass by and curse thy fill—but pass, and stay not here thy gait.'
> (17.66–74)

The epitaph preserves the two separate versions represented in Plutarch, the first Timon's own, the second penned by the poet Callimachus. In retaining both, the text suppresses and presents the dead man's name with almost the same breath. Again, editors have tried to explain the contradiction in terms of the play's unfinished state. But what's notable is the fact that the play's starkest inconsistencies should consistently arise at the site of the grave and do so in a form that concerns the play's root aspiration as a historical document: the seeking out of Timon's name from its epitaphic ur-text, the ambiguous point at which Timon becomes history, or at which history becomes fiction.

What is the commonality between the two accounts? In both cases, the epitaphs seem to address their own unstable rhetorical condition. If in the first an inability to read seems nevertheless to produce a reading, in the latter the wish to pass by, to avoid the confrontation with the name, is linked to its nevertheless having been brought forth. Both epitaphs in that sense take the form of an interpretive snare that recalls the coils of "apprehension": Whether forcibly grasping or foreclosing, one renews an inscriptive effect constituted out of such repeated divisions, losing what one would apprehend, conjuring what one would forgo. That is how an impression "takes" whether we want it to or not: The wax "interprets for my poor ignorance," the soldier says, as if it worked of its own accord. Inclusive exclusion, the measure of sovereign potency as well as the structural condition of the play as a coherent form, now assumes the more radical form of a mad iteration at the textual root of the play, an undoing of autopoiesis as a coherent mechanism. The sense of contagious expansion attending a convergence between poiesis and aesthesis, text and its beyond, is intimated in the way the epitaph's referential deixis seems to swallow up the world beyond it, undoing the distinction between auto- and heteroreference at the core of autopoietic formation. Having just spoken of taking the "character" in wax, the soldier concludes with reference to Alcibiades: "Before proud Athens he's set down by this, / Whose fall the mark of his ambition is" (16.9–10). "Set down by this": by this time before the gates of Athens, or set down by *this*, by the wax and, by extension, the texts—this text—within which his story is inscribed? Whose "fall" is it? Textual self-reference and thematic heteroreference unstably converge in the instituting deixis.

"Some beast read this," Timon's epitaph declares, recalling the scene at the gates where he likens citizens to wolves. Faced with the disarticulations of textual space and origins provoked by Timon's buried/unburied condition, the play ends with a scene that attempts to assimilate the unruly epitaph by gathering it into a final negotiated settlement staged at the now secured city walls, and in doing so to reestablish itself as a coherent signifying system. The appearance of the city's citizens on the tiring house makes apparent that the boundary between audience and play is once again at issue, though now in the form of a human parlay. Where Timon had called on the wall to "dive into the earth," the citizens now are called upon to "descend," leaving the walls intact. As they descend, the report of Timon's death arrives and the epitaph is read, transported thus from "the very hem o'th' sea" to the now humanized limits of the polis. Whatever untoward possibilities implied by the thought of a beastly and nonhuman form of reading are banished.

What is at stake in that negotiation between Alcibiades and Athens is nomos understood specifically as a matter of spatial bounds. To confirm his good faith, Alcibiades promises that with his entry to the city not one of his soldiers "shall pass his quarter"—leave his allotted space—"or offend the stream / Of regular justice in your city's bounds / But shall be remedied to your public laws / At heaviest answer" (17.60–63). The arbitrary force of the inclusive exclusion that founds the polis is now translated into the rationalized though no less arbitrary "cull[ing]" necessary to appease the appetite of the aggrieved: "By decimation and a tithed death," the Second Senator tells Alcibiades, "If thy revenges hunger for that food / Which nature loathes, take thou the destined tenth, / And by the hazard of the spotted die / Let die the spotted" (17.31–35). The hunger that "nature loathes"—that is, the cannibalistic consuming of one's own that has driven appetite in the play—is answered by the measurable calculus of a biopolitical "decimation," the killing off of one in ten. Destiny—the inhuman dimension compelling the narrative—becomes the calculated randomness of thrown dice. It's not hard to hear in that enumerative language the ameliorative sacrifices of the rationalized state to come, a construction whose forced nature is evident enough in the tonal queasiness of the final punning reduction of death to a gaming die.

The "confusion" that had been the unstable grounds of the polis—the transgression of limits as the condition of limits—is redressed now by way of a temporally ordering signifier. "Throw thy glove," the Athenian senator requests of Alcibiades,

> Or any token of thy honour else,
> That thy wilt use the wars as thy redress,
> And not as our confusion, all thy powers
> Shall make their harbor in our town till we
> Have sealed thy full desire.
>
> (17.49–54)

With the promissory token, reiterative division is converted to deferral, a recourse secured precisely insofar as it is put off. That anticipated resolution hangs, ultimately, on the seal of memory. Having translated the sea-lapped gravesite—uncanny figure for a mourning that goes on without us—into an irenic image of weeping for "faults forgiven," Alcibiades concludes by announcing his entry into the city even as he consigns Timon to future remembrance: "Dead / Is noble Timon, of whose memory / Hereafter more. Bring me into your city" (80–82). Alcibiades anticipates the story we now receive—we are gathered into the final settlement—but only by suppressing the reiterative force of the rhetorical inscription the play has just placed before us. Timon is buried in memory, retained insofar he is passed by, held inwardly by virtue of the foreclosure that makes such assimilative internalization possible, whether we think of that incorporation in relation to the consolidated subject or polis.

In one regard, Alcibiades's memorialization appropriates Timon's life to political ends; in another, it annuls the political altogether. The final invocation of memory is consistent with a turn at the end from reiterative division and a mourning that continues with none by to a subject-centered, intentionalist understanding of history that now absorbs everything, including the city's walls:

These walls of ours
Were not erected by their hands from whom
You have received your grief; nor are they such
That these great tow'rs, trophies, and schools should fall
For private faults in them.

(17.22–26)

What had constituted the very possibility of narratable political and subjective space is now gathered into a story of faults neatly distributed according to a distinction between private and public, present and past, death now no more than the mitigating arbiter of such distinctions: "Nor are they living / Who were the motives that you first went out" (17.26–27). Associated in its incorporative vocation with such a historicist understanding of history—history as causal "beads on a rosary" in Benjamin's memorable phrase—memory stakes itself against the more incalculable force of material inscription, history as what "takes."

By foregrounding aestheticization as the necessary condition of political and subjective foundation in the era of exchange, as what makes its groundless ground knowable, *Timon of Athens* also exposes the instability at the heart of that consolidating process. Reproducing the split condition of its textual sources, conveying forgetfulness—passing on—as the condition for memorialization, the play's various, multiplying epitaphs locate history ultimately neither as a matter of securely attributable faults nor as a throw of the die but as the more fundamental contingency that opens the very distinction between intention and mere chance. Are the "confusions" of Timon's death meaningful or a matter of textual accident? The play is political precisely insofar as it compels decision by virtue of the irresolvability of that distinction between what is phenomenally sensible and what remains inassimilably arbitrary, an uncertainty that can't be accommodated by aesthetic consciousness and thus one that exceeds "apprehension" in all its ambiguous extension from aesthetic unknowing knowingness to the supposed "capture" entailed in sovereign self-founding.

Leaving such distinctions undecidable, the drama situates us within history's contingent effects rather than granting us the role of history's transcendent surveyors, refusing us even the comfort of dire tales of sovereign origination such as Agamben's. Confirming us as torn and temporally disjunct beings, the play—what is "set down by this"—in its circularly posited and positing condition makes us responsible for a choice which at the same time will have already been made insofar as it constitutes us as conscious, intending or indicative subjects. Returning to the question of the conditions of sociality as expressed in the play's organizing opposition between gregarious bonding and misanthropic solitude, we can say that the drama leaves us at once autonomous and bound, neither within nor beyond the pale, misanthropolitans all. At the boundary of its world-destroying animus, the play thus intimates the possibility of a social economy understood, not in terms of a dialectical interplay between individual freedom and submission to a collective totality but as a domain in which every element "takes," becomes itself by yielding itself, gives what it doesn't possess.

3

A New Air

Tragedy, Being, and History in Anthony and Cleopatra

In her compelling analysis of *Anthony and Cleopatra*'s relation to the possibility of thinking the world beyond the totalizing terms of imperial mastery, Jane Hwang Degenhardt discovers the play's promise in its capacity to express "possibilities—or indeed worlds—that are left open and alive through an embrace of not knowing."[1] For Degenhardt, that prospect of a generative acknowledgment of limits arises in the play and in the era via an emerging conception of the horizon. Employed as a means of conceiving the world immanently, absent a secure transcendental vantage, the horizon during early modernity orders the world even as it makes evident the relativity of human knowledge. In doing so, the horizon becomes "a tool that celebrates the virtues of what we cannot know—allowing us to embrace the unknown as a source of potential and an opening for alternative understandings of world and of history."[2] Conceiving the world in such immanent terms means embracing it in the present moment of one's contingent perception, to grasp the world not through Caesarian "reason, rational planning, and the spatial and temporal structures of empire" but as Cleopatra does, by "being alive to the ever-present possibilities of now."[3]

Reading such an account in the present context, one is struck by how closely such a formulation of resistant epistemology aligns with the lineaments of tragic knowledge as it was evoked during the era, a knowing unknowingness underwritten by the confrontation with limits. Indeed, Degenhardt's formulation may provide an insight into the kind of ideological work such a version of finitude is capable of performing. To the extent that we acknowledge the contingency of the subject as well as of the world, we need to ask whether the horizon is what the subject conceives in the urgency of its present experience or whether it is the horizon that confirms the subject and world alike in their mutually reinforcing opposition. Even if we imagine it as a matter of unrealized possibilities, the world conceived in the perspectival terms Degenhardt proposes will remain conceptually a world laid out before the subject's gaze and thus a world that remains within the terms of a certain form of sovereign mastery. As we noted in relation to the ironies of posthumanism, insofar as they conceive the world in such representational terms, as descriptively before us, even those accounts that purport to conceive a world in which the human subject is no more than one incidental element among others inevitably reinstate the claim to a form of speculative command.

Anthony and Cleopatra does indeed trouble Caesarian conceptions of self and world, I will argue, but precisely insofar as it resists such a representational conception. A work in which the fateful event—Cleopatra's betrayal—repeats itself often enough to leave us unsure where we are in the story, in which death coincides with erotic fulfillment or passes with the lightness of a thought, and where the staginess of the close leaves it unclear whether the end occurred then or is occurring here and now: If anything, to experience *Anthony and Cleopatra* is to experience the dissolution of every structuring horizon and the undoing of any confident sense of the nowness of the now.[4]

In one regard, that erosion of the fateful event as event in *Anthony and Cleopatra*—everything that suggests we are experiencing the tragic undoing of tragedy—merely writes large the logic of the missed encounter, recognition as missed recognition, that has always sustained tragedy's power to ensnare; it is that force of ensnarement that aligns tragedy with autopoiesis and thus with the sense of being embedded in a world that will always exceed us. Still, something in the logic of tragedy shifts here, as if at the juncture between the tragedies and romances Shakespeare glimpsed the possibility of another mode. "Com'st thou ... from / The world's great snare uncaught," Cleopatra asks Anthony toward the close.[5] Such an escape is possible—and it is possible only insofar as it remains at the level of the question—to the extent that the play unbinds autopoiesis itself as a structuring mechanism. In a work in which tragic agonism is internalized as betrayal and where betrayal is also beguilement, where tragedy's fatal web is indistinguishably the net of erotic seduction, equating the law of finitude with the law of desire, tragic pathos turns into the system-dissolving force of the death drive. With that, the grounds of tragedy itself shift from an epistemological concern—Hegel's as well as Luhmann's framework—to a question of being, the enigmatic dimension toward which Cleopatra, the figure "whom everything becomes," insistently draws it.

Rather than escaping finitude in an affirmation of the "now" of human experience, that shift from cognition to being inscribes the fact of finitude and history more deeply. With the undoing of the event-character of drama, the missed encounter of tragedy reveals being itself as sheer, iterative contingency, as the fact of its historicity. *Anthony and Cleopatra*'s alterity, its resistant promise as woman-centered and anti-imperial tragedy, is a function of the reconception of space and time attending that formal transformation. Rather than enabling perspectival consciousness via an awareness of spatial and temporal horizons, the play realizes the force of history and reference as precisely that which traumatically blinds the speculative eye of reason insofar as it undoes the frame by which consciousness sustains itself in a posture of knowing unknowingness in relation to the world. Unbinding the organizing horizon between past and present, the play scuttles the narrative logic by which historicist history—history "according to the victors," as Walter Benjamin puts it—is structured. Instead of discovering the promise an unrealized past holds for the present, an oppositional construction that preserves the present as "ever-present," the play realizes our own historicity as the inscribing force of what we will have missed, other worlds in what will have already displaced us in the moment of our apprehending them. Rather than an object of inquiry, being, in that sense, proves indissociable from its own questioning, with all that implies of absolute proximity and distance.

The possibility that in *Anthony and Cleopatra* tragedy itself will be betrayed is evinced from the outset in the play's systematic undoing of the agonistic structures that usually propel tragic drama. An outward assault that inwardly wounds, betrayal—the drama's central preoccupation—unbinds the distinction between self and other even as it undoes the opposition between romance and martial aggression. With Cleopatra's turning tail at Actium, Anthony's forces "kiss . . . away" his fortune (3.10.7). Indistinguishable from beguilement, the betrayal that marks the fall colludes with the desire it betrays. "I will betray / Tawny-fine fishes," Cleopatra the angler says, evoking her technique of erotic ensnarement (2.5.11–12). Nor can we be certain that Actium itself, the fateful event of imperial history, wasn't a scene of seduction rather than war, the faithful realization of a love pact committed from the outset to "let Rome in Tiber melt" (1.1.35). "You knews't too well / My heart was to thy rudder tied" Anthony says, describing his disastrous flight (3.11.55–56).

That slackening of drama's clarifying tensions is bound up with the play's reimagining of identificatory structures more broadly. The difference between Rome and Egypt, male and female in the drama is not an opposition: It is the difference between opposition and that which exceeds such structuring logic. Thus for instance, at the level of self-relation, the distinction between Anthony's manly, botched, and specularly conceived suicide—"there is left us

/ Ourselves to end ourselves," Anthony declares; "a Roman, by a Roman / . . . vanquished"—and Cleopatra's gentle, oppositionless erotic dissolve, as well as her claim to find in death a place beyond winning or losing (4.15.21–22, 4.16.59–60). Associated with male rivalry, Roman face-to-face confrontation is equally aligned with familiar structures of specular identification and empathy. "When such a spacious mirror is set before him" as Anthony is in death, Roman Caesar weeps—"Caesar is touched" (5.1.33): that the most calculative of characters should be the most lachrymose is no contradiction, bearing out as both do the specular character of Roman thought. In her elusive, unmarkable death, Cleopatra would slip the frame of such mimesis, refusing the emperor's objectifying triumphal pageant—mirror of his triumph—and the audience's punctual claim on her, the forms of empathetic identification by which we would make her ours.[6] At the close, we're left with little more than the affectively queasy trace of a figure that may have slipped away long before the play ever began: "This is an aspic's trail; and these fig-leaves / Have slime upon them, such as th'aspic leaves / Upon the caves of Nile" (5.2.349–51).

Cleopatra eludes such oppositions because the difference she embodies or enacts is constitutive of them, a paradox familiar enough from French feminist thought.[7] How that contradiction—of a piece with the queen's power to "make hungry where she most satisfies"—bears on the play's representational status can be teased out from one of the play's more obvious riddles. On the one hand, none of Shakespeare's plays to this point is as explicit about its own mediated condition. The play is shot through with intruding messengers, a fact that undoes a confident sense of temporal sequence. And it is a play in which characters seem already acutely conscious of their own historical reception—their "place i'the'story"—whether in the form of the famous lover's self-conscious staginess—"The nobleness of life / Is to do thus [*embracing Cleopatra*], when such a mutual pair / . . . can do't"—or Caesar's whisking us off in the heat of the moment to look over the written record that he believes will absolve him in the eyes of history (3.13.46, 1.1.38–40, 5.1.73–76).[8] As Janet Adelman has pointed out, just about every scene in the drama is framed, with characters offering commentaries on the action as it unfolds.[9] "Behold and see," incidental Philo belittlingly declares by way of introducing our protagonists at the play's opening (1.1.13). Escaping that mastering regard and finally alone, the famous couple seek out their most intimate pleasure by turning their gaze on others: "all alone / Tonight we'll wander through the streets, and note / The qualities of people" (1.1.54–56). The charismatic pair at the core of the drama—the object of view that draws us—merely offers back the image of our own voyeuristic investments.

On the other hand, divided and mediated through and through, limitlessly relational as it would seem to be, the drama is no less distinctive for its preoccupation with sheer, immanent being or becoming. Drunken Lepidus's capacity to be taken in by Anthony's account of the exotic crocodile being "shaped like itself" is repeated in the play's broader, less comically resolvable fascination with self-creative self-identity, a fascination that crystallizes in the figure "whom everything becomes" and who becomes everything (1.1.51).[10]

The contradiction is only apparent, for in *Anthony and Cleopatra* mediatory division is immanent being, an aporetic condition evident in the passage that amounts to the romance plot's primal scene. We receive the description of Cleopatra on her barge on the Cydnus as Shakespeare's reiteration of North's translation of Plutarch's account of an event described by someone who didn't witness it. Such layering displacements are inseparable from the scene's erotic power, reiterating through to its intimately coreless core. As critics have pointed out, Cleopatra is described entirely through her circumambient adornments, the object itself amounting to a scene of an indeterminately auto- and heteroerotic captivation in which the distinction between negation and affirmation, chastening and eroticization, dissolves.[11]

> The poop was beaten gold,
> Purple the sails, and so perfumed that
> The winds were lovesick with them; the oars were silver,
> Which to the tune of flutes kept stroke, and made
> The water which they beat to follow faster,
> As amorous of their strokes . . . . . .
> . . . . on each side her
> Stood pretty, dimpled boys, like smiling Cupids,
> With divers-coloured fans, whose wind did seem
> To glow the delicate cheeks which they did cool,
> And what they undid did.
>
> (2.2.199–212)

Affirmed in its denial, erotic desire proves at once self-sufficient and inescapable. Evoking a desire that is produced in the movement of its division, the scene undoes the distinction between erotic object and its mediating, reflexive apprehension. "Her gentlewomen, like the Nereides, / So many mermaids, tended her i'th'eyes, / And made their bends adornings" (2.2.213–15): To turn your eyes attentively on her eyes is to become adornments of the scene itself. Cleopatra herself becomes an instance of "O'er picturing ... Venus," exceeding the goddess and eluding us in the self-same movement by which she becomes, or fails to become, picturable, thus undoing the observer/observed opposition around which Luhmann's systems theory is organized.[12]

With that indistinction between object and its witnessing, desire aligns with a division that remains inapprehensible insofar as it precedes and determines the grounds of phenomenal experience as such—a gap in nature: "The city cast / Her people out upon her; and Anthony, / Enthroned i'th'market-place, did sit alone, / Whistling to th'air, which but for vacancy / Had gone to gaze on Cleopatra too, / And made a gap in Nature" (2.2.220–24).

Understood in these terms, *Anthony and Cleopatra* would seem to articulate a turn in Shakespeare's conception of theatrical representation. John Gillies observes that the play's joking on the tautology of self-identity—Anthony's clarifying observation that the crocodile is like itself, etc.—may have been prompted by the explanatory recourse to such comparatives in contemporary travel literature, accounts seeking ways to make sense of alien worlds.[13] We can expand the claim. The play's thematic concern with alterity—the possibility of female-centric tragedy, and the limits of Roman, which is to say, Western, universalism—is intimately bound up with a larger shift from mimetically conceived drama—theater as mirror of an already constituted or proposed world—to theatrical figuration as phenomenally constitutive. The play anticipates the romances and the formal paradoxes marking their representational world, their mixture of self-conscious artifice and a heightened commitment to the immanent "sounds and sweet airs" of the human and inhuman sensorium.[14] That paradox arises because sense certainty is itself an effect of the reflex through which phenomenal being at once comes to and divides itself, conjuring itself as pure artifice—as theater.

Such a turn, and its inevitable opening beyond the phenomenal dimension it enables, is apparent in the play's peculiar, and again romance-like, drift toward allegory in its later stages, where a character named Scarrus comes to bear the inscribing wounds that define him—"I had a wound here that was like a T / but now 'tis made an H"—and where a soldier named Eros oversees Anthony's final embrace of erotic fatality, as well as in the recourse to forms of negational conjuring: Anthony's god, Hercules, first appearing at the point we hear him depart (a turning away of the gods that for Heidegger defines tragedy as such); the repeated references toward the close to "discandying," suggesting a sugared artifice whose dissolution nevertheless undoes the world.[15]

The play would *seem* to mark such a formal evolution: In fact, it's not hard to find at least traces of all of those elements in Shakespeare's earlier plays.[16] That uncertainty is to the point. As against Caesar's triumphally progressive version of history, Cleopatra's recursive conjuring—what Alain Badiou terms the torsional "ultra-one" that determines the possibility of any form of accounting—is precisely what resists efforts to locate it in developmental terms, to account for it via narratives of origination.[17] And yet, precisely in unbinding such a continuist account of history, Cleopatra's autopoietic creationism at the same time draws the play closer to the fact of its historicity. To understand that paradox, I want to turn to the scene that is often taken to be the pivot-point of the tragedy, as well as the founding event of Western history according to the epic tradition. Notably, the battle at Actium is not shown: We hear of it through the framing commentary of the witnesses. Moreover, for those who do see it, the effect is blinding—"To see't mine eyes are blasted ... Mine eyes did sicken at the sight"—and self-violative—"I never saw an action of such shame—/ Experience, manhood, honour, ne'er before / Did violate so itself" (3.10.3, 15, 21–23).

The dire consequence of such witnessing is bound up with the status of the action, its character as history. As critics have noted, the tragedy is staked in the end on the question of how its events will be told. With her suicide, Cleopatra explicitly seeks to elude her inscription within Caesar's intended triumphal procession—a figure for the lineal, successive conception of imperial narrative history—and thus within history as a narrative of the victors; she slips from his gaze as she slips from ours.[18] Actium amounted to the keystone of that Caesarian narrative construction. An end, the event equally marked an origin, the beginning of Octavius's imperial consolidation and of Roman history as the era of "universal peace," and thus as a prefiguration of the Christian era.[19] Providing that linear narrative the formal boundary that shapes it—that lets it become a narrative at all—the epic event is more than an event in history: It is "the triumph of history itself."[20]

To turn tail at the climactic moment, "i'th'midst o'th'fight / When vantage like a pair of twins appeared, / Both as the same," is to betray the narrative as a narrative, deflating the agonisms that propel the drama, just as Cleopatra will at the tragedy's close seek out a space beyond the logic of winning and losing (3.10.11–13). Staged as a scene in which all is gathered into a suspended moment of specular reflection "when half to half the world opposed," Actium represents the drama's climax as an enactment of its own condition as a mimetic form—a "spacious mirror" of identificatory rivalry (3.13.9, 5.1.34). In that regard, the scene crystallizes the play's status as exemplary history—its inscription within the atemporal, tautologically self-affirming circuits that make up Roman universal history, a history in which Caesar's victory at Actium reflects Aeneas's founding victory, which reflected Caesar's victory at Actium as already prophetically figured within the circle of Aeneas's shield. "Swerving" from "the great face of war," just as Anthony had been dismissively seen to "now bend, now turn" his devotion toward her, Cleopatra performs the self-violative reflex that undermines from within such specular oppositions, undoing sight, manhood, and reason in the process (3.11.49, 3.13.5, 1.1.4).

The effect of that betrayal of mimesis is realized as much in the scene's reception as it is in the scene itself, dissolving that very distinction. As Adelman points out, with this pivotal moment, "the frame has replaced the picture," a condition of "o'er pictur[ing]" that consigns the audience—subject and object—to the same vacant "gazing" as Anthony's abandoned navy (3.13.12).[21] In a play in which knowledge and identity is constituted relationally, as the "behold and see" of spectatorial power, to be deprived at such a climactic point of a narrative object is to have our gaze return upon us, to be made conscious our own lurid, self-confirming investments in such moments of resolving violence. We too are robbed of the face-to-face of the mimetic encounter and in the process experience the shame that marks all within the scene: "I never saw an action of such shame"; "He's unqualitied with very shame"; "'Twas a shame no less / Than was his loss to course your flying flags, / And leave his navy gazing" (3.10.21, 3.11.43, 3.13.10–12). Oppositional difference is replaced by internal division, and because that self-difference is constitutive, shame's action can't be separated from that to which it responds. "O wither has thou led me, Egypt? See / How I convey my shame out of thine eyes / By looking back what I have left behind / 'Stroyed in dishonor," Anthony declares, shamefully avoiding his shame (3.11.50–53): Such reiterative recoils merely repeat Cleopatra's illegible "action," an action (or "Actium") of seduction and of abandonment, of fearful aversion and of drawing closer, of origin and end. Actium converts the manly face-to-face of exemplary imitation and epic history into a sourceless contagion of blind following that, undoing the distinction between object and reception, extends beyond the play's frame. "Thou knew'st too well … thou shouldst tow me after," Anthony says, a line that equally describes the affective undertow of tragedy itself (3.11.55–57).[22]

Even as it undoes history as a mimetic form, however, that state of originative following suggests the deeper grounds of historicity as such, the condition that leaves all, like Anthony, "lated in the world" (3.11.3); it's not difficult to hear in Anthony's "looking back what I have behind" something like the forming condition of historical consciousness.[23] The suspended nature of the moment of face-to-face confrontation bears out the purely formal character of mimetic drama; as imperial epic, the victory at Actium amounts to a victory over time itself, Quint observes.[24] The breaking open of such a narrative construction—the intrusion of "that which within history interrupts history"—amounts to the breaking in of time, even as it institutes reference—a referring beyond—in that breaching of the closed system of mimetic tropes.[25]

In the epic tradition, Cleopatra's flight coincides with a deictic interruption that untethers the narrative's temporal logic. "The queen herself was seen to call upon the winds / Spread her sails, and now, even now (*iam iamque*), to let loose the slackening canvas," runs Virgil's description of Aeneas's prophetic shield.[26] Tasso describes the scene as it is depicted on the sculpted portals of Armida's palace: "And now behold—the outcome of the battle as yet inclines to neither side—behold the barbarian queen in flight."[27] The contingency of the moment—the sharp *now* of temporal experience—is felt in the breaching of narrative time, a break produced at once through and against the layered mediations from which it emerges: inscriptions on an ekphrastic shield and the forms of a sculpted gateway. That the "now, even now" of deixis should also blind is not merely a function of where one's sympathies lie. It arises because reference as such is instituted with the undoing of speculative cognition, the singularity of historical indication—"the way in which the 'there' is disclosed for existence," as Heidegger puts it—arising precisely in the breaking of the abstract tropic equivalences that constitute speculative reason.[28]

The question of *Anthony and Cleopatra*'s relation to history comes to the fore toward the close in the play's novel articulation of tragic finitude. Whereas William Junker reads the play's critical historiographic opposition as that between the factical history of the imperial conqueror and Cleopatra's messianic fulfillment of an unrealized past, Emily Vasiliauskas argues that the play is structured around an opposition between an imperial, supercessive history in which the past, and indeed death itself, is sublated into a timeless, artifactual present and a recursive aesthetic figured in Cleopatra, an aesthetic that, disrupting such sequential narration, opens the possibility of conveying finitude—death itself—as a "pure phenomenal present."[29] For Vasiliauskas, the very possibility of representing finitude, and thus of constituting secular tragedy as an aesthetic form, hinges on that disruption of the historiography of the victor.

Though I'll want to question the opposition between mere artifice and phenomenal presence that governs her conclusions, Vasiliauskas is right to suggest that the possibility of representing tragic finitude in the drama hinges on a disruption of imperial historiography, history as supercessive narrative sequence—on the possibility, that is, of an alternative temporality spoken from the margins. In what sense does Caesar's linear, narrative form of accounting fail history? Not because it restricts itself to what has actually occurred, nor because it imposes an artifice on the rich phenomenal domain beyond it, but because it presumes a transcendental posture outside historical time, the posture from which one claims to know the narrative chain as a chain, the same posture by which we claim to know such an account as merely fictive or artifactual or false. What would it mean, then, to conceive history from within the movement of historical temporality, without transcendental recourse? To see that, we need to follow Cleopatra's rigorously devious lead.

Anthony and Cleopatra is not the only Shakespeare tragedy to involve a falsified death and to conclude with two suicides rather than a single fatal denouement. In *Romeo and Juliet* as well tragic recognition, such as it is, amounts to a sequence of missed encounters.[30] But orchestrated from within by the character who suffers its consequence, such a device in the later tragedy becomes something more than an expedient for the doubling required by romantic tragedy, something that fundamentally transforms the status of the tragic work and its relation to historical causality. The tragic conclusion precipitates from the point at which Cleopatra sends the contrived account of her death to Anthony, who then in response falteringly kills himself. Cleopatra's behavior might be taken as an especially damning instance of her flighty narcissism—"Hence, Mardian, / And bring me how he takes my death," she declares—or simply as a panic reaction: in either case, a spontaneous, unforeseeing act (4.14.9–10). The fact that it is her waiting woman who first broaches the plan emphasizes the seat-of-the-pants feeling of the moment. On the other hand, it can be construed as the moment where Cleopatra commandeers the play, assuming authorship of her own history—after all, she will indeed kill herself. Either she acts from within the narrative moment or from a controlling posture beyond it, from within or without the story.

And yet, neither of those alternatives recognizes the more radical form of knowing that the scene intimates. "She had a prophesying fear / Of what hath come to pass," Diomedes later tells the dying Anthony (4.15.121–22). A prophesying fear after or before the fact of her fatal message? A fear that brings about what it fearfully foresees? "Foresee me … a good fortune," Charmian demands of the soothsayer, dissolving the distinction between will and fate (1.2.14–16).[31] An exorbitant being—"madam, keep yourself within yourself," Charmian says—one given over to transports and proleptic forebodings, Cleopatra can be seen actively to institute the reverting and rivening knowledge that will come to characterize the play's end and that defines tragedy as tragedy (2.5.76). "Caesar, thy thoughts / Touch their effects in this. Thyself art coming / To see performed the dreaded act which thou / So sought'st to hinder / … you are too sure an augurer," Dolabella tells Caesar as he arrives too late before the final tableau (5.2.327–32). Insofar as we allow the possibility that Cleopatra is constituted through the recursive autopoiesis she initiates, insofar as we grant her the capacity to act at once from within and beyond her contrivance, authoring the fiction within which she is at the same time inscribed, we unbind the imperial causality that would designate her deceptions as *mere* deceptions. In that regard, the phenomenal event as such, indeed, the fatal event that marks the passage into history, is conjured out of a recursive spiral of pure artifice—a fabricated message answering a message that had answered it. Such a real and unreal, phenomenal and aphenomenal condition is symptomized in the ambiguous status of the deaths that mark the close, as weighty as Anthony's awkwardly hoisted body or, for Enobarbus and Iras, as ethereal as a thought or a kiss, where true death is no different than a death on stage.

Understood in these terms, the tragedy introduces a version of time that anticipates the radically dialectical, finite history forwarded by Walter Benjamin in "Theses on the Philosophy of History." In terms that strikingly align with the play's opposition between Roman time—the universalizing, supercessive "history of the victors," as Benjamin puts it—and the elliptical time Cleopatra introduces, Benjamin counters historicism's "continuist" version of history as causal sequence—time as an accretion of discrete instances like "beads on a rosary"—with a nontranscendent history constituted purely from within the dialectic of time's relational configuration, one in which the moment assumes its meaning solely by virtue of its answering relation to a prior or future instance: thus for Benjamin the "weak messianism" of the relation to futurity—"our coming was expected on earth"—and the "tiger's leap into the past" by which the present takes on meaning. "Having been arises from the future," Heidegger writes in a comparable set of reflections on historical immanence.[32]

As it does for Shakespeare, such a conception implies a paradoxical understanding of the moment. Given the distinction between historicism's chain of discrete moments and the constellating relationality implied by dialectical history, one would assume that it is the former that would be associated with a valorization of immediacy. And yet for Benjamin materialist history alone opens the moment as a "time filled by the presence of the now [*Jetzizeit*]," a monadic "time of the now ... blasted out of the continuum of history."[33] For William Junker, who anticipates me in recognizing the connections between Benjamin's essay and Cleopatra's counterhistory, Benjamin's "Now time," realized in Cleopatra's theatricalized romantic consummation at play's end, involves a "confluence" or "unifying isomorphism" of an unrealized, nonfactical past and the present within a "dilated 'Now'" that "preserves and at the same time dissolves" temporal differences even as it anticipates a messianic fulfillment, becoming a "sacrament of ... future restoration."[34] But such a coalescent and sacerdotal account, with its echoes of Hegelian dialectical sublation, risks losing sight of the more radical character of Benjamin's equivalence between dialectical disruption and nowness. If materialist history is associated with the now, it is because it is the breaking open of continuist temporality, with its universalizing conception of time as a series of formally equatable terms, that manifests time in its sheer, nonrelational singularity. It is the disjunctive conjunction of the dialectic that constitutes the moment in the first instance—history, again, as the "event" of its undoing. The circularity of such a conception of history as that which exceeds its theorization suggests the necessarily performative, implicative, and self-cancelling nature of every fully dialectical account, including Benjamin's. For that reason, the manly "blasting out" of the historical continuum Benjamin celebrates can never be purely distinct from the eye-blasting shame of Actium, that which "traumatically blinds any unprepared faculty of cognition," Werner Hamacher writes, describing Benjamin's "now" time.[35]

The unbinding of sequence implied by Cleopatra's precipitating message will continue toward play's close in the lovers' curiously asynchronous endings, each seeming to "o'er take" and follow the other's end.[36] Eroding the boundary between past and present, life and death, that recursive, lapping movement ultimately extends beyond the limits of the stage and comes to bear on our own status as witnesses. Immediately after the defeat at Actium, beguiled again by his "grave charm," Anthony answers Virgil's invocation of him, though now in a context of headlong precipitancy rather than stabilizing mimesis.

> I will o'ertake thee, Cleopatra . . .
> Eros!—I come, my queen.—Eros!—Stay for me.
> Where souls do couch on flowers we'll hand in hand,
> And with our sprightly port make the ghosts gaze.
> Dido and her Aeneas shall want troops,
> And all the haunt be ours. Come, Eros, Eros!
>
> (4.15.44–54)

Virgil's appropriation of Actium to seal Aeneas's imperial aims is here answered immediately after the fateful defeat by Anthony's appropriation of Virgil to affirm his romantic apotheosis. It's a misreading, of course—Aeneas's epic ambitions are premised on his abandonment of Dido; she will bitterly refuse his advances in the underworld. Then again, to reprove Anthony thus may involve its own error: Historically, Anthony precedes Virgil; it might well be Virgil who self-interestedly distorts the founding tale. But then, such legitimation according to proximity to historical origins relies on its own no less forced narrative construction.

Ultimately, textual "authenticity" in such a context will be entirely a function of whether we align ourselves with the claims of empire or romance. For the audience, the questionable allusion prompts a moment of heightened interpretive self-consciousness, a sense of implication picked up in Anthony's reference to the gazers within the scene. It's a disquieting detail insofar as it suggests a reversal: If we are figured there, it's not as living witnesses to theatrical ghosts from the past but on the side of the dead witnessing their "sprightly" caprices. In a play in which characterological self-consciousness increasingly becomes a matter of seeing oneself as if already dead—think of Enobarbus at the point of death invoking history's witness—the audience, too, here finds itself on both sides of the line. Such a crossing is, in fact, an indissociable feature of interpretive consciousness. Beyond questions of political or personal affiliation, interpretive force arises at such moments of textual undecidability ultimately because one's condition as living and agential is staked on knowing where fiction ends, on determining the boundary between artifice and an animate beyond. And yet it is precisely such an always defensive assertion of interpretive force that implicates, leaving us again and from the outset, insofar as we are conscious and intractably historical beings, ghostly onlookers to our own passage.[37]

Such an inscriptive effect recurs more explicitly toward the drama's close. Facing the prospect of being displayed in Caesar's triumphal pageant, Cleopatra famously forewarns Iras that "Mechanic slaves / With greasy aprons, rules, and hammers shall / Uplift us to the view . . . and I shall see / Some squeaking Cleopatra boy my greatness / I'th'posture of a whore" (5.2.209–11, 219–21). Vasiliauskas argues that the metatheatrical scene evokes the queen's elusive power precisely insofar as such a description so obviously falls short of representing her.[38] Catherine Belsey similarly argues that it presumes Cleopatra as an "idea" beyond the textual representation we now witness.[39] I would argue that Cleopatra is never more present than she is at this moment precisely in her empty, irrefusable theatricality. For one feels in such a moment that she has foreseen us, that, often out ahead of herself, the exorbitant queen "oe'r takes" us as well, catching us in the recursive loop of her ceaseless self-theatricalization. Because her beguiling presence was always a function of its self-negating turns, she is all the more present "now, even now" precisely in the uncanny vacancy of the metatheatrical moment.

Nor is the disquiet easily allayed. To construe such a perturbation as a theatrical effect, to treat theater as mere spectacle before our Caesarian gaze via a confident assumption of the distinction between artifice and what lies beyond it, is to renew its opaque force, even while bearing out the truth of the shaming accusation the passage levels against us: Our "very force entangles / Itself with strength," as Anthony's entangles Anthony (4.15.48–49). Cleopatra eludes us, even as she eludes Caesar, and in doing so irrefutably returns precisely in the redounding failure of our mastering apprehension. The form of temporal ellipsis that had marked the lovers' tragically missed encounters is here reiterated as the condition of theatrical representation as such: Even as Cleopatra, anticipating us, exceeds her time and the limits of the stage, the audience finds itself drawn back within the diegetic scene. Or will have been: To know the effect is already to have missed it. Undoing the distinction between distance and proximity, retreat and approach, theater reveals the grounds of its phenomenalizing power as a conjuration that takes place strictly speaking nowhere—neither then nor now but in the chiasmus of those temporal markers' crossing. And yet, out of the contingency of that missed encounter—what breaks open continuity and mimetic logic alike—the blinding now of historical reference takes place.[40]

Reinhart Koselleck has written of the significant shift from *historia magistra vitae*—the essentially atemporal formal constructions of exemplary history—to a modern conception of progressive history, the collectively singular understanding of historical time as its own determining force.[41] Covering similar ground, François Hartog articulates the emergence of a sense of future-oriented historicity during the eighteenth century, associating it with encounters with the fact of cultural contingency: Hartog focuses on Chateaubriand's New World travels.[42] Reflecting similarly on the limits of imperial, universalizing historiographic modes, *Anthony and Cleopatra* adumbrates some such reconception of historical temporality, though if anything in a more radical form. The progressive history whose emergence Koselleck describes is a developmental construct however open-ended that development remains. The paradoxical character of such a progressive conception as "a totality . . . but a totality that can never be complete" is a function of its speculative character, its status as "a history that becomes history only by its passage through consciousness."[43] From that which suspends time in exemplary history, mimesis becomes that which produces time as progression by virtue of its sublation within the consolidating movement of inward reflection.[44]

The more thoroughgoing dissolution of such conceptions in Shakespeare's play—everything that makes it a precocious anticipation of modern critiques of history as historical consciousness—occurs because it approaches time specifically in relation to the problem of mimesis and its own status as theatrical representation. Revealing tragedy's formal syncope as the source of its female heroine's being and power, the play translates the missed encounter of tragedy into a thoroughgoing questioning of the grounds of mimetic identification and thus a posing of the question of the relation between theater and historical reference as such. A quasi-mythical figure, Cleopatra "herself" is something like a real theatrical being. The uncanniness of her stage presence is of a piece with the disquiet one would feel coming across an accoutrement of hers in a museum: not so much a demystifying proof that she was historical as a confirmation of the reality of her irreality—the momentary experience of a gap in nature. History as such takes such a form, the play suggests: a speculative caesura, the return of that which wasn't there in the first place, being as the fact of its elliptical historicity.

Paradoxically, it is theater that has the power to convey historical time as that speculative wounding. With its emphasis on the formative "behold and see" of theatrical relation, the play evokes not an appropriative mimesis of time whether in its exemplary or developmental form but a mimetic disjunction within time, the self-positing division by which temporality as such is instituted.[45] More fundamental than the constituting temporal interval Luhmann sees as the core of reflexive autopoietic formation, an account that presupposes time as a medium, such primordial spatialization—Benjamin's "dialectics at a standstill"[46]—implies a "transcendence within immanence," that which exceeds time's continuum insofar as it constitutes it.[47]

Anthony and Cleopatra transports tragedy through and beyond the logic of autopoiesis as Luhmann frames it, as infinite and open as that logic may be. Extending the missed encounter of tragedy absolutely, the play enacts the dehiscence of reflexive formation as prior to, because formative of, the speculative terms in which Luhmann conceives system formation, however implicated those categories of observer and observed are for him, and prior to because constitutive of the organizing temporality such a speculative conception implies. Breaking open the mimetic conception of theater and history, the play aligns the missed encounter of tragedy with the blind force of reference, including the reference by which one would designate such an autopoietic effect as "there," in the play or as a knowable system. Constitutive of subject and world alike, that referential force—what undoes even as it opens the possibility of mimesis—amounts to the infraction of historical contingency, one that registers as what in the now of our living apprehension will have already situated us elsewhere, on the side of the dead, a dislocation renewed in the act of its articulation.

In the context of tragedy, that reiterative force takes the form of an eye-blasting fall of speculative reason, an undoing of the world as knowable. On the other hand, insofar as it figures that effect as being itself, *Anthony and Cleopatra* intimates within the very movement of that dislodgment the opening of another, unthought world. Having evoked toward the close her dream of an extravagant, world-striding Anthony, Cleopatra addresses her audience,

Dolabella, a figure whose tender skepticism might be taken as the embodiment of the audience's own tragic knowingness:

Cleopatra:	Think you there was, or might be such a man As this I dreamt of?
Dolabella:	Gentle madam, no.
Cleopatra:	You lie up to the hearing of the gods! But, if there be, or ever were one such, It's past the size of dreaming. Nature wants stuff To vie strange forms with Fancy, yet, t'imagine An Anthony were nature's piece 'gainst Fancy, Condemning shadows quite.

(5.2.93–100)

What Cleopatra would evoke in "one such" is not the Anthony she has just described; it's past the size of even her dreaming and beyond the play as mimetic representation. If it is available at all, it is only via the self-undoing spiral of propositional logic: the stuff of fancy beyond nature and of nature beyond fancy, including *that* fancy. Such a one exceeds the opposition between fancy and nature, subjective and objective, inside and outside, willed and unwilled, insofar as it involves the ellipsis that opens such distinctions in the first place, implying a sizeless space absent the frame or horizon by which we conceive what we take to be the world.

It *would* be that if it were, and the possibility of such another world depends on the strange imperative conditionality of Cleopatra's exchange. "You lie," she tells Dolabella, with the self-certainty of a truth claim. But what she asserts as the truth holds to the form of a conditional: "But if there be, or ever were one such." What she imagines beyond the terms of truth and imagination takes the form of something like an absolute conditional, the hypothetical suspension that opens the space of truth and falsity, real and unreal. "Have you escaped the world's snare unscathed?" Cleopatra had asked, and it remains a question, for to claim otherwise, to claim a posture beyond tragedy's hold, would be to renew an autopoietic mechanism initiated via such a presumption to outsideness. But another prospect, neither subjective nor objective, locatable neither in the self nor in the play, is intimated in the interval between, the interval by which the question opens as a question. To think world otherwise than as a mechanism of autopoietic capture or a space within which being is inscribed, to think being itself as "being-with," as sheer relationality, is to think being as the condition of always already responding that Heidegger posits as "thrownness" and the play enacts from the "Nay, but ..." with which it opens to the "nay, but" by which, within and against implacable tragic certainty, Cleopatra harbors the possibility of "such a one" at the close.[48]

Benjamin sensed the promise of such an alternative version of space and time within Shakespeare's tragic catastrophes. In *One-Way Street* he writes:

> Again and again, in Shakespeare, in Calderón, battles fill the last act, and kings, princes, attendants and followers "enter, fleeing." The moment in which they become visible to spectators brings them to a standstill. The flight of the *dramatis personae* is arrested by the stage. Their entry into the visual field of nonparticipating and truly impartial persons allows the harassed to draw breath, bathes them in new air. The appearance on stage of those who enter "fleeing" takes from this its hidden meaning. Our reading of this formula is imbued with expectation of a place, a light, a footlight glare, in which our flight through life may be likewise sheltered in the presence of onlooking strangers.[49]

Benjamin's "dialectics at a standstill" recurs here as the suspended condition of those who "enter, fleeing," a coming/going that compresses to its limit the temporal suspension of theatrical representation itself, that which according to the "Theses on the Philosophy of History" "flashes up at a moment of danger" here corresponding to what is fleetingly caught in the glare of the stage.[50] Such an aporetic mode of appearing—an emergent fleeing corresponding to the space between within which all are held at the close of the play—implies as well an elusive form of rapport, one in which distance—the glare of the footlight and the gaze of a stranger from another time—amounts to a shelter, exposure, a space of intimacy.

The plight and the promise of those who "enter, fleeing" has, of course, particular resonance for a historical moment in which a large portion of the world's population is consigned to such a suspended condition. The rethinking of relation and world demanded by such a state is, for Benjamin, ineluctably bound to the temporality it implies. For the "new air" breathed within that space of irreducible exposure is what Benjamin in the "Theses" calls the "open air of history," history as dialectics "at a stand-still" whose truth is inseparable from its phantasmatic character, as what, like Cleopatra, appears in its disappearing. "The true historicity of historical objects lies in their *irrealis*. Their un-reality is the store-place of the historically possible. For their *irrealis* indicates a direction through which that which could have been is referred to those *for whom* it could have been and for whom it is preserved as a—missed—possibility."[51] The promise of unrealized historical possibilities is not what appears across the horizon of past or future. Such a framing—the world as representation—is what via a specular configuration of self and world translates the world into the "spacious mirror" of our own conceiving: no world at all. Rather, history is precisely what in flight breaks that frame, opening the self in the rupturing noncoincidence of its own generative finitude. Rather than representing what is or might be, a mode that inevitably posits the world as there to be represented, such originative noncoincidence constitutes world in its formative and fugitive irreality, as the thrownness by which being exposes itself as what is absolutely near and far, a world beyond us insofar as it is at the same time intimately ours.

4

Comedy and Romance

The Autopoietics of Socius and World

Absent the considerable philosophical tradition establishing tragedy's pertinence for issues of subjectivity, history, and sense experience, comedy and romance as genres remain for the most part just that, literary genres, which means that it is impossible to engage the relation between those forms and autopoiesis without confronting the sticky question of genre's legitimacy as a category of historical analysis. The very fact that Shakespeare scholarship came to speak of "Problem Comedies" is indication enough that as a category comedy is as murky for us as genre is in Shakespeare generally: It is no more intuitively obvious that *Troilus and Cressida* or *The Merchant of Venice* should be comedies than it is that *The Tragedy of King Richard II* should be a history. And, while it was certainly a literary designation, "romance" doesn't become a category for drama until the nineteenth century, and when it does, it's a genre defined largely by its promiscuity. "The cipher we call 'romance' is almost a symbol of the inadequacy of 'classic' generic classifications."[1] Little wonder that recent criticism tends to forgo the historically uncertain middle distance of genre for apparently firmer empirical or methodical ground, focusing on concerns that cut across generic distinctions: the material conditions of the text, rhetoric and poetics, acting practices, sociological themes and issues that would seem to bypass the question of literary mediation altogether.

But ignoring genre in all its historical and epistemological fuzziness may amount to losing sight of the work itself. As Fredrick Jameson points out, "genre" describes literature under the aegis of its social legibility, as something like a social contract constituting the terms within which the work should be received.[2] In that regard, the uncertainty of comedy as a category may be precisely to the point. That Shakespeare comedies should have titles such as

As You Like It and *What You Will* may be indication enough of the genre's self-conscious relation to the inextricability of meaning and reception and of the indeterminate consequences of acknowledging that equivalence. Comedy, it can be argued, is the genre that poses the question of genre and, insofar as what we will or like is the play tout court, of the literary as such. Comedy realizes its vocation as a distinctly social form by revealing that literature is nothing other than the contract by which it is received, even while leaving the nature of that contract as undecidable as our capacity ever to know what it is in literature that we will or like.[3] To put this another way, Shakespearean comedy and romance play a privileged role in articulating the emerging phenomenon of aesthetic experience in all its ambiguous social and political consequence during the era.

To focus on such aesthetic matters is to swim against the tide of critical trends that have sought to move away from a formal literary focus to a consideration of the determining force of material context: staging practices, repertory history, those facts of mixed authorship that tease us out of any conception of the work as an autonomous phenomenon. And yet, to undertake such a shift in the name of contextualization is to presume in advance that we know what context is and, in doing so, to fail to recognize the larger stakes of the aesthetic as it was emerging during the era. For it is through the problem of the aesthetic and, indeed, through the question of the work as an autonomous form that the era engaged the relation between work and world as a problem, posing the question of how such a distinction is constituted in the first place. Paradoxically, it is precisely through a consideration of literary autonomy that we engage the political stakes of literature and literature's relation to world formation during the period most profoundly.

The logic of that association between comedy and literariness can be understood in terms of the genre's relation to autopoiesis, in all its human and inhuman dimensions. As we saw, in tragedy, finitude takes the form of a division at the heart of anagnorisis itself—recognition as missed recognition—an effect that aligns tragic pathos with the inhumanly mediating force of techne, including the techne of theater. Aestheticization plays a role in tragedy, as *Timon* makes obvious, in relation to the effort to make that necessarily inapprehensible autopoietic ground knowable; it is the inevitable failure of that process of phenomenalization that bears out tragedy's insistent reference beyond itself as an aesthetic form, a referential drive apparent in tragedy's distinctive relation to history, including in the case of *Antony and Cleopatra* the historicity of being as such.

In comedy, autopoietic division is redoubled in the reflexive formation of the work as a self-conscious aesthetic totality, a sublimating process that translates the reiterative effects of tragic finitude into a generative relation between prohibition and desire. The insistence with which the romantic comedies, especially through their epilogues, at once frame themselves and bypass their framing as literary forms reveals the genre's explicitly autopoietic logic, as well as its intimate relationship to emerging conceptions of aesthetic consciousness. The genre's status as a distinctly social form is directly related to that explicitly aesthetic character: through its limitlessly self-framing enactments, Shakespearean comedy demonstrates its power to comprehend its own purely mediated condition and thus to establish a relation to the fact of relationality itself as the grounds of meaning. Comedy is a distinctly social genre not because it reflects society but because it bears on the paradoxical conditions of social relationality as such.

The autopoietic nature of that ambition becomes especially apparent in the romances, where the aesthetic work's constitutive exorbitancy, its moving beyond itself to become itself, is fully realized in the work's extension beyond death's limit. In absorbing that formative limit—what constitutes the distinction between art and world—the romance form would incorporate all into its totalizing aesthetic ambit. The consequence of that absorptive ambition is apparent in one of the genre's more striking formal paradoxes: The most self-consciously artful and artificed of Shakespeare's works, the romances are also the most committed to an apparently unmediated evocation of the sensory world as such. Because that convergence is a function of the self-divisive, self-incorporative movement by which aestheticization surpasses human limits, such an effect is most fully realized as an evocation of a sensible world beyond us—the ambient "sounds and sweet airs" of an island with no human witness by. The late works suggest, in that regard, the essentially aesthetic underpinnings of the more affirmative variants of vital materialism and posthumanism.

At the same time, in thus explicitly swallowing its own determining formal limit, aesthetic fulfillment in the late works harbors an annulling force more thoroughgoing than tragic pathos, a blind mechanical pulsation at the core of vitality itself that undoes sense, self, and world altogether. Whereas tragedy stakes itself on the finite limit between work and world and in doing so transgresses that boundary, in absorbing that finitude into its omnivorous aesthetic form, Shakespearean romance at its extreme threatens to render the very distinction between spirit and matter, art and lifeless techne illegible. In political terms, traced to such a limit, aestheticization can tilt either in the direction of fictions of "embodied social thought,"[4] a vitalist conception whose darkest variant would be the superhuman authority of the leader's living voice, or toward as yet unforeseen versions of relational justice beyond every form of aesthetic or social totalization.

1

To bear out these claims for comedy's distinctively political-aesthetic stakes, it's helpful to rehearse the familiar—perhaps over-familiar—story about the derivation of the dominant version of the genre, among other things in order to register Shakespeare's significant formal innovation, one that amounts to conceiving comedy as a self-consciously autopoietic form. Whereas "Old Comedy" of the classical tradition had a directly referential satirical aim, "New Comedy"—the form observed by most English Renaissance practitioners—defined itself in terms of its own more or less dialectical formal structures, a developmental scheme based on resistance, trial, and reconciliation. Thus while comedy was an inherently social form—a matter of human relationality rather than, say, depth subjectivity—it was also a manifestly autonomous form, one associated particularly in the Renaissance era with a heightened degree of reflexive literary self-consciousness.

The paradox is only apparent, a matter of expanding our sense of what autonomy means. Romantic comedy in particular, the argument runs, observes a familiar if infinitely variable pattern: We move from an initial conflict with a harsh or irrational law, usually associated with a paternal authority, to an interval of transgressive license—a passage into the topsy-turvy, identity-dissolvative space of an Arden or a fairy wood—concluding with a reengagement of social law, yielding a new, more flexible society or at least a reconsolidation of identities and some form of intergenerational reconciliation. Modern theories of Shakespeare comedy tended to be elaborations of this formal structure, with Northrop Frye, for instance, aligning the process of trial and rejuvenation with mythic and natural cycles and C. L. Barber viewing the structure as a movement of restraint leading to saturnalian release and "clarification," that is, the "clarification about limits which comes from going beyond the limit."[5]

That last formulation suggests the epistemological stakes of the structure, comedy becoming a matter of reflectively internalizing and thus cognizing law's arbitrary, founding bar, translating that limit into "an inner source of coherence," as Frye puts it, a meliorative internalization of law consistent with comedy's broadly humanistic vocation as a social structure.[6] Because that intractable limit—what aligns law with finitude—is also the defining limit of fiction's own status as play, such an internalizing structure equally realizes comedy's character as a reflexive form, its capacity to reflectively compass its own constitutive bounds. Whereas satire in its referential claims presumes the distinction between the drama and what lies beyond it, New Comedy in its incorporative relation to law stages the conditions for that very distinction between work and world, fiction and its beyond, a fact evident enough in the playful porousness of its endings.

Shakespeare heightens and problematizes such claims for the self-conscious literariness of the form and does so by virtue of an insight that undoes comic structure in the developmental form proposed by Frye and Barber, even while it lies at the heart of Shakespeare's remarkable purchase for modern psychoanalytic theory. In Shakespeare, law and what transgresses it turn out to be indissociable, a fact that troubles law and desire alike: Desire can't do without the bar that constitutes it as desire, even as law, instituted by way of its transgression, realizes itself precisely through its negation and deferral—the "tarry a little" by which, as we will see, Portia at once, or by turns, suspends and enforces judgment. Such complicity, the hallmark of an explicitly iterative autopoietic turn, is evident from the outset in Shakespeare, for example in the cartoonish theatricality with which the young lovers embrace their woe in *A Midsummer Night's Dream*—"Oh, cross! Too high to be enthralled to low"—or the outright masochism with which love seeks out what thwarts it: "Herein mean I to enrich my pain, / To have his sight thither and back again," says Helen of her jilting lover, setting the plot in motion.[7]

The peculiar impression at such moments that the characters of comedy are at once puppets buffeted by the course of events and preternaturally conscious of those inscribing structures, Helena's "thither and back again" articulating the Barberian theory of comic loss and reconsolidation to a tee, reflects the way in which the inextricability of law and transgression undoes comedy's developmental process of internalizing cognition. The characterological "deepening" that occurs with Shakespeare's subsequent comedies merely deepens the epistemological contradiction. The motivational obscurity of Rosalind's impulse to maintain her forest disguise even after she has encountered her new love is in no way resolved as the love play "evolves." The apparent sadism with which Rosalind as Ganymede as Rosalind undertakes to "cure" her beloved of love—"men have died from time to time, and worms have eaten them, but not for love"—perhaps in the name of "testing" that love might lead one to wonder if she is in love at all. Yet the moment Orlando vacates the scene and Celia confronts her, she dissolves in a tizzy: "O coz, coz, coz, my pretty little coz, that thou didst know how many fathoms deep I am in love!" (*As You Like It*, 4.1.92–94, 179–80). Although such contradictions can be psychologized—read as Rosalind's self-protective relation to her own desire, for instance, or a countering desire for power—they are ultimately simply the defining mark of desire as such, which is to say, desire insofar as it enacts its own definingly "barful strife" (*Twelfth Night*, 1.4.40). According to such a logic, the more proximate the love, the more distant it becomes; the more distant, the more intensely present. "The poor world is almost six thousand years old, and in all this time there was not any man died in his own person, *videlicet* in a love cause," Rosalind playing Rosalind declares: In Shakespeare, desire as such is always vicarious, by definition never yours at all: "What is yours to bestow is not yours to reserve," Caesario declares (*As You Like It*, 4.1.82–85; *Twelfth Night*, 1.5.175).

Instead of being either a psychological or a social form, Shakespeare's comedy enacts the autopoietic constitution of that distinction between inward and outward, self and socius. Rather than merely presupposing desire's founding bar, *Twelfth Night*—Shakespeare's most conceptually ambitious comedy—enacts the enigmatic origins of sociality itself. At play's start, Olivia and Orsino figure the two inoperative poles of social being: the one, Orsino, narcissistically self-referential, enlisting others to "act [his] woes" (1.4.25); the other, Olivia, radically other-directed, given over to interminable mourning for a dead brother that amounts to a devotion to the intractable otherness of death itself. And yet, absent the bar that would relationally constitute them, the two extreme identities are no identities at all, committed as both are to a pursuit that undoes the very distinction between autonomy and heteronomy, life and death. Whereas Olivia submits to her daily rounds of tears "to season / A brother's dead love, which she would keep fresh / And lasting in her sad remembrance"—a matter of keeping death itself forever alive—Orsino in his self-absorption finds himself, Actaeon-like, subject to desires that, "like fell and cruel hounds, / E'er since pursue [him]," a condition of limitlessly self-completing self-division (1.1.29–31, 21–22).

Any possibility of self or world in such circumstances will be a function of a life-giving caesarian intervention, the relation-forming go-between who in dividing constitutes: Caesario's role at the start of her play is comparable to Rosalind's at the close of hers, where, as epilogue, she "insinuates" herself between the men and women of the audience in order to constitute both as desiring subjects, which is to say, subjects at all. At the same time, that intervention constitutes Caesario as a subject. When she speaks of assuming "such disguise as haply shall become / The form of my intent," "become" means *become* as well as correspond with, as her purpose will be a precipitate of the assumption of the mediating role that determines it and sets the play in motion (1.2.53–54); a function of the fiction's constructions, "Viola" is not named until the end of the play, and even there in a conjectural form, as a figure forever awaiting the recovery of the "maid's weeds" that will let her assume her proper persona.

At play's opening, emerging from an unseen tempest, we only know that she mourns her apparently drowned twin, a condition that amounts to a form of self-mourning: As Caesario, she will mime the deceased brother's identity.

And yet that divided assumption—the condition of any identity at all—will depend on her entering the structuring terms of the relations she precipitates. The rapidity with which she vacillates between inquiries about the Duke to identificatory investment in Olivia, who similarly mourns a brother, back to a decision to serve Orsino in his pursuit of Olivia suggests the formative character of the triangulated rapport on which the rapid-fire sequence lights, the self constituting itself not through identification with another but via a prior identification with the relational difference that constitutes any possibility of identity; desire begins for all, in that sense, as the desire of the desire of the other.[8] The go-between constitutes the structure within which she intervenes and constitutes herself in the process; that Viola imagines she will assume such an identity as a eunuch, implying the severing loss of what was never there in the first place, merely bears out the constitutive function of division in whatever social and subjective being it inaugurates. Caesario in that sense enacts at the opening what Hymen enacts at the close of the earlier comedy, Hymen being the god in whom barring partition is indistinguishable from consummation, who resolves by redundantly barring what bars us—"I bar confusion"—and whose appearance the audience too can't help but experience as a resolving breach, a deus ex machina where none, we thought, was called for.

Understood in these formative terms, Caesario amounts to a figure for theater itself, that other mediatory go-between that at once intervenes in and creates worlds. Or, she is a figure for the audience, we whose interpretive intervention—a traumatic breach in its own right—constitute the meanings that at the same time captivate us. The terms of the triangulation can be rotated indefinitely, for they simply manifest comedy's founding truth of irreducible relationality, a sourceless relationality that constitutes the terms that make it up.

The explicitly literary nature of such an illimitable condition is evident in the insistently metatheatrical aspect of Shakespearean comedy, although it's a literariness founded precisely in our inability to confidently designate it as such. It is no coincidence that at the climactic moment when he has finally achieved his erotic prize with the choice of caskets, even with the flesh-and-blood Portia there at hand, Bassanio oddly lingers over the painted portrait, that suspension being the condition of desire as such in the play. Rather than being sublimated in Barberesque fashion, the prohibitory law of the father—what "bars [Portia] the right of voluntary choosing"—is merely renewed in the "sweet ... bar" of the portrait's queasily "severed lips," such dehiscence—the untethering force of figuration as such—now become what it always was: desire's animating spur (*As You Like It*, 2.1.16; 3.2.128–29).

The founding nature of such a passage into the register of art—the fact that it constitutes the desirous beings it captivates—makes it as hard to locate the end of the effect as it is to find its origins. "The sight of lovers feedeth those in love," declares Rosalind-as-Ganymede, as she "removes" to observe in the encounter between Silvius and Phoebe the "pageant" of "true love" "truly played": "Bring us to this sight and you shall say / I'll prove a busy actor in their play" (3.4.46–53). And she will: Chastising Phoebe, she makes herself by virtue of such admonitory checks the very object of the desire. "'Ods my little life, / I think she means to tangle my eyes too!" (3.5.43–44). But that inciting bar is implicit simply in the posture of exteriority she assumes as onlooking analyst of love's pageant: The more one presumes a place outside the scene of love, the more fully inscribed within—an enactment of the logic of autopoietic formation as erotic condition. The audience will be subject to the same effect at play's end, where it will be precisely our desire to mark the close of a thoroughly iffy drama and thus our place securely beyond it that will have inscribed us within it. Presuming we know what "Rosalind's" "if I were a woman" means, presuming it implies a securing and in this instance manly truth behind fiction's conditionality, we are caught out when s/he solicitously curtsies at the close, betraying the truth of our desire against our desire, its barful truth in that againstness. "Your only peacemaker," Touchstone declares, such iffyness is also the condition of identity as such in the comedies.

The most developed such scene of inscription involves Malvolio's master class in the perils of close reading. The very arbitrariness of the signifiers that the forged love letter places before him tempts him to the exercise of interpretive force that will confirm his mastery: "And yet to crush this a little, it would bow to me, for every one of these letters are in my name" (*Twelfth Night*, 2.5.123–24). Recalling the bow demanded by an audience at play's close, such demarcating force narcissistically reflects—"If I could make that resemble something in me"—insofar as it bears out Malvolio's power to reduce the letter to the cognizable object of his gaze (107–8). But because the interpreter as autonomous subject is indeed constituted by way of such violence, Malvolian reading inevitably amounts to an act of self-mutilation. Possessing the letter, the letter divides and possesses him, inscribing itself within: The next scene finds Malvolio speaking the script verbatim. "Thou hast put him in such a dream that, when the image of it leaves him, he must run mad" (2.5.168–69). Nor is it clear how the dream ends. Reiterating the negation implicit in the boundary-drawing force that makes it signifiable, the letter's call for him to prove himself by "be[ing] opposite" to all appeals means every check will only deepen its hold.

It is not just Malvolio who is autopoietically captured thus. "Possess us," Malvolio's maligners declare, asking Maria to tell them of the planned device.

And the onlookers will indeed be at least as possessed as the victim. Looking on from the safe obstruction of the boxwood tree, the all-seeing audience is as lost to the ploy as Malvolio:

> Sir Toby: Fire and brimstone!
> Fabian: O peace, peace.
> Malvolio: And then to have the humor of state and, after a demure travail of regard, telling them I know my place as I would they should do theirs, to ask for my kinsman Toby—
> Sir Toby: Bolts and shackles!
>
> (2.5.45–50)

Riven as the figure he observes, Sir Toby suffers the "Do it! Don't do it!" masochism of sitcom spectatorship, a matter again of being inscribed within the division that would secure you; it is precisely the presumption of spectatorial distance that creates the identificatory lure, a condition that leaves no one in the scene able to "know [their] place." "*If this fall into thy hand, revolve,*" the letter instructs (2.5.125). The term means "reflect," but one can imagine the comic stage business attending the phrase, with Malvolio turning behind himself to see and not see his audience within or perhaps even beyond the stage. For the seeing makes no difference: All within and beyond the stage are inscribed in a reiterative effect predicated on the claim to know, where the "revolve" of reflective self-consciousness is ever the turning back on oneself that divides and blinds. To determine that M.O.A.I signifies the *real* truth that "I am O"—nothing, a "naywood"—or to vigilantly acknowledge that all the characters are anagrammatical variants of one another does nothing to preserve us from the blindly inscribing enactment of what such knowledge presumes to know.

That performative unknowing knowing extends from the smallest instance of interpretive enactment to the epistemological space of Shakespearean comedy as a whole. Bassanio is never blinder than he is when he is transfixed by Portia's portrait, lost to the moment precisely when the end is in sight. And yet he himself describes that as an artistic effect, and we can equally say that Bassanio—and the play—is never more knowing than at this moment: Portia herself is, after all, no more than a fiction, a portrait. In other words, the scene figures the knowing and absorbed condition of all theatrical experience, the madness of our own inside-outside postures as readers and spectators.

In Shakespeare, the "thither and back again" of comedy amounts to the movement by which fiction self-incorporatively reflects back on itself in order to comprehend itself *as* fiction, even as in that turning it divides and exceeds itself: Thus Shakespearean comedy's distinctive bounded unboundedness, which is to say, its status as an aesthetic formation. The "reality effect" of the form, indeed, the illusion of a boundary between fiction and its beyond, is a function of that recursive self-seeing: "If this were played upon a stage now, I could condemn it as an improbable fiction," says *Twelfth Night*'s Fabian (3.4.115–16). Understood thus, comedy as a genre is neither an objective trait of the work nor a function of subjective response. Its "law" consists in the movement—the "revolve"—that ceaselessly constitutes and reconstitutes the domains of subject and object, fictive and real in the first instance. Shakespearean comedy, in that sense, is the privileged genre for an articulation of the Montaignean performative knowing unknowingness—knowledge as the knowledge that one does not know—that characterizes aesthetic consciousness as a historical phenomenon. That transformation—the signal emergence of aesthetics as consciousness—is directly related to the explicitness with which, registering the indissociability of desire and prohibition, inclusion and exclusion, Shakespeare realizes sociality itself as an autopoietic condition.

For Shakespeare, comedy's ethical claims are inextricable from that literary effect. The personally and socially consolidating aims of the exclusions that tend to mark comedy's close—the casting out of the Shylocks, Malvolios, and Calibans—are evident enough: Maria, for one, wins her wished-for husband by virtue of her devising the trick that leads to Malvolio's imprisonment. What makes such a scapegoating distinctively Shakespearean is its visibility. The fact that embittered Malvolio holds the key to the play's resolution, Viola's female clothing remaining in the hands of the sea Captain who is now imprisoned by virtue of Malvolio's suit, makes clear that in victimizing the other in the name of social and subjective consolidation one inevitably victimizes—suspends and divides—oneself.

For Shakespeare, as for Montaigne, the acknowledgment of such contingency amounts to a supervening ethic. If Hymen affronts by intervening in a play that was well on its way to a seamless resolution, we will have ourselves to blame for that: We'd been told that Ganymede was raised by a magician in the forest. Our having construed that as a mere fiction is simply the measure of our absorption in the tale we have been told, bearing out our inability to hold onto the fact of there always being "more woeful pageants than the scene / Wherein we play," as Duke Senior tells Jacques (2.7.137–38). To acknowledge the irreducible taint of that exclusionary violence and what it tells of a domain always beyond our ken is to recognize the fact of one's status as a social being; to refuse it is to renew the violence.

Although such a posture will be foreclosed with the harsher gender politics of the tragedies, as it will by the religious agonisms of the era to come, it would not be a mistake to see in the kind of performative knowledge Shakespearean comedy articulates—a knowing founded on the knowledge that one does not know—as the humanistic precursor to Enlightenment social conceptions and indeed to contemporary variants of social-construction theory based on an acknowledgment of the contingency and situatedness of all identity claims. Given that, it is all the more important to recognize the aesthetic foundations of such a positioning and to acknowledge that aestheticization involves its own forms of force, a violence less apprehensible and more untoward than the exclusions aesthetic consciousness acknowledges as its own. Whatever its self-exceeding character, the recursive "revolve" of aesthetic consciousness remains a speculative and, ultimately, a consolidating mechanism: Interpretive mastery redounds upon the self, but it is the self against which it redounds, and while the turns may be illimitable, such a performative conception retains the sense that it is the self that sets them in motion. But comedy intimates another force outside the speculative return on nonreturning of aesthetic consciousness.

That force is bound up with a contradiction that marks aesthetic logic at its core. Comedy's claim—indeed fiction's claim—as a limitable category of illimitability writes large an aporia built into identity formation in the plays. We noted the necessity of Viola's identification at play's start with the difference that constitutes any possibility of identity; the suspended, inside/outside posture distinctive to aesthetic subjectivity depends from the outset on the possibility of reflectively cognizing one's divided condition. And yet such an identification with the conditions for identification is obviously contradictory and unstable, depending as it does on a radically arbitrary and unmotivated positing in advance of the subject who would make it signifiable. Paradoxically, aestheticization's claim to self-referential autonomy merely exacerbates the problem of reference as such. Such a force beyond the aesthetic logic that consolidates self and play as phenomenally coherent forms is intimated both in the unrepresented traumatic loss that precipitates *Twelfth Night*'s opening and in the play's vexed engagement with law at the close, where the finite boundary that determines indeterminate fiction as fiction devolves into a mad indistinction between containing and "enlarging," suppressing and delivering, describing and becoming mad—where reading madness becomes, for Feste, a mad reading.

The instability of comedy's efforts to sustain itself as an inclusively autonomous aesthetic form, and thus the uncertainty of its capacity to serve as a model for social relation, is most apparent in that relation between art and art's limit term, law, a fact evident enough from the Shakespeare comedy that resolves itself—or attempts to do so—in a court of law. If the case of Martin Guerre reveals the intrusion of the aesthetic at the heart of law's workings,

The Merchant of Venice shows a relation to law as the internal condition of aesthetic formation.

Portia invokes the law in two forms at the close. She enjoins the force of judgment against Shylock by turning the letter of the bond back against him. "My deeds upon my head! I crave the law" (*The Merchant of Venice*, 4.1.203). Then, subsequently, she invokes the criminal law against any alien seeking the life of a citizen by "direct or indirect attempts," an interdict that leaves Shylock confiscate of his goods and life (4.1.346). The two laws are discrete, though intertwined: The fact of having failed at his civil case would seem to leave Shylock exposed to the criminal charge—the first leads to the second—and yet the second law against direct or indirect means would seem to compass and supersede the prior bond.

The strategic logic of that doubling of judgment is clear. Holding Shylock to the terms of his bond guarantees the moral authority of a redounding judgment: The condemned has no recourse because he has drawn the punishment upon himself. Indeed, one can argue that the translation of what the play takes to be the Judaic literality of law into Christian spirit consists of nothing more than this reflexive, internalizing movement of the letter upon itself, imitating the structure of conscience. With that, authority preserves its purity even as it effects judgment: I have merely given you what you ask for, you have answered yourself. And yet such a logic remains vulnerable to endless *tu quoque* reversals. Thus the demand for an externally imposed judgment, a fixed and autonomous law, and indeed a law that determines the boundary between internal and external, citizen and alien. But on its own, such a law would leave itself exposed to the taint of its own enactment: The more fiercely the alien is expunged, the more that exclusion becomes the defining mark of the state, the more that exclusion will haunt, as Shylock may well haunt the play after his banishment. Law fulfills itself, then, insofar as it maintains both of those forms *almost* as if they were one. The peril arises, I'll suggest, insofar as they are in fact one.

Much is at stake in that conjoined and divided law beyond its practical effectiveness. Across its two terms—the self-authored and the extrinsically determined—the entire range of often reversible binaries that structure the play can be mapped: internal vs. external, spirit vs. letter, Christian vs. Jewish, ethical vs. political, civil vs. state. On a different scale, Portia's doubled judgment embodies a contradiction at the heart of modern Western conceptions of law itself and its relation to culpability. On the one hand, guilt is conceived in relation to the intentional actions of a knowing subject. On the other, ignorance of the law is no excuse. The tension is endemic to a social dispensation founded at once on claims of individual autonomy and the universality of state and law, a contradiction that would have stood in sharper relief during the pre-Westphalian moment in which such categories were consolidating.[9]

Such a contradiction is more than a contingency of the duplicitous ends that law must strategically serve; it is the condition of law as such as an explicitly autopoietic form and, indeed, the very the source of its force as law. The at once mutually defining and mutually interfering relation between aesthetics and law crystallizes at the moment when Shylock holds the knife to Antonio's chest, tempting us to invoke the force of law directly. The recoiling consequences of such a resolving response become apparent at the point of Portia's riveting suspension of that denouement, her "tarry a little" just as Shylock touches Antonio's breast with the sharpened knife (4.1.302). Shylock is on trial in the moment, brought to death's limit and thus given the chance to relinquish his hold, or given the rope with which to hang himself. Antonio too is placed on trial, tested to see if he will abrogate the fierce hold of his martyrdom. But Portia tests herself as well here, letting herself feel the temptation to eliminate the rival. "To bait fish withal," Shylock says of the object of his apparently gratuitous desire for the pound of flesh, and it's not hard to see how the punitive act functions as a lure (3.1.50). For it is precisely in eliminating the other that one renews the other's hold, revived in the form of inwardly redounding guilt. That Portia grasps the logic is borne out when she inverts it by enlisting Antonio as go-between in the final, triangulating resealing of the ring bond: If by excluding one ends up reviving the other's grip, by including one disables the other; Antonio's presence will, in fact, be muted from this point on.[10]

The audience, too, is made to feel the snare of that violent exaction, the lure that would turn comedy to tragedy. It is impossible not to experience a sense of dramatic deflation in Portia's suspension of the climactic moment. That even the presumptively Christian audience would be subject to a degree of undifferentiatedly sadistic pleasure at the prospect of Antonio's martyrdom is evident enough in the tonal instability of the response of Antonio's loving compatriots to the news of his losses:

Solanio: Now, what news on the Rialto?
Salarino: Why, yet it lives there unchecked that Antonio
hath a ship of rich lading wrecked on the narrow
seas—the Goodwins I think they call the place—a
very dangerous flat and fatal, where the carcasses of
many a tall ship lie buried, as they say, if my gossip
Report be an honest woman of her word.
Solanio: I would she were as lying a gossip in that as ever

knapped ginger or made her neighbors believe she wept for the death of a third husband. But it is true, without any slips of prolixity or crossing the plain highway of talk, that the good Antonio, the honest Antonio—O that I had a title good enough to keep his name company—

Salarino: Come, the full stop.

Solanio: Ha! What say'st thou? Why, the end is he hath lost a ship.

(3.1.1–16)

The florid attenuation of the Venetian citizen's recital is indication enough of the appetite for violence that propels unchecked rumor. At other points, the good citizens will bait Shylock to revive his animus: "But tell us, do you hear whether Antonio have had any loss at sea or no? . . . Why I am sure, if he forfeit, thou wilt not take his flesh. What's that good for?" (3.1.39–40, 47–48). That same communal autoaggression recurs in Tubal's insistently returning Shylock to the most excruciating of his own losses: "Thou torturest me, Tubal" (3.1.113).

Resisting that reiterative violence, and thus establishing the possibility of community at all, is an aesthetic matter. For Portia's "tarry a little," that which suspends the act, even as it opens the space for the audience's self-conscious reflection on its own impulses, amounts to a mobilization of the resources of the aesthetic, an enactment of the power of art's suspension of purposive aims. "I pray you, tarry," Portia had said to hold Bassanio off from making his casket choice, lingering out its exquisite torture (3.2.1): Theater itself is just such a call to tarry a moment. The aesthetic retains that salvific power by virtue of its defining opposition to law as supervening, inscriptive forms. The agency and knowingness one claims in tarrying depends on the sense of such restraint as a suspending interval in relation to a prospective endpoint, even if that end is indefinitely deferred. At the same time, positive law assumes its punctual force as against the groundlessness of fiction. It is by virtue of their apparent opposition that fiction and law assume whatever instrumental claims they hold; through it, we can comprehend our absorption as a merely fictitious effect—aesthetics becomes aesthetic consciousness—even while law can claim its fixative power as absolute and as its own.

But that distinction, on which much depends, is itself a fragile fiction. "Tarry a little," Portia had said, suspending law, holding it from the space of the play. "Tarry, Jew, / The law hath yet another hold on you," she repeats, reinvoking law. What risks exposure in that reiterative, mutually opposed and mutually interpolated relation between law and fiction is the constitutive force of the tarrying itself, the fact that the suspension that defers law is ultimately indistinguishable from the bar that institutes it, as it institutes the aesthetic, in the self-divisive movement by which each takes hold. In other words, law obeys its own autopoietic logic. Constituted in the movement of its division and deferral, law reveals its sourcelessness as the very condition of its inscriptive force, realizing itself as that which in the moment of ones responding—the moment of ones refusal—will have been already in force, already determined. The two variants of law the play enlists against Shylock—the dialectically reactive and the extrinsically fixed—merely play out the irreducible contradiction at the heart of law's force. As at the court of Artigat, law as inescapably missed encounter reveals itself as an autopoietic effect, which, undoing the event of its own founding, derives its inscriptive force from its effacement of the distinction between singular and universal, response and object, inward and outward; the law is mine alone, addressed solely to me, even as it precedes me and grasps me in advance—thus its power to mark intimately and infinitely.[11] It's a condition fulfilled at the close in Shylock's enforced "conversion," a final withholding of law that inscribes law more deeply and primordially as an inward and alien division.

And yet, insofar as law too proves a limitless autopoietic effect, the "tarry" of pure aesthetic consciousness, absent the securing opposition to finite law that defines it, becomes instead a lingering life as afterlife, a taint that continues on beyond the drama's tidying end—"I pray you, give me leave to go from hence. / I am not well," says Shylock as he exits the play—and one that was in fact present from its start. "In sooth, I know not why I am so sad," Antonio begins the play,

> It wearies me, you say it wearies you;
> But how I caught it, found it, or came by it,
> What stuff 'tis made of, whereof it is born,
> I am to learn;
> And such a want-wit sadness makes of me
> That I have much ado to know myself.
>
> (1.1 1–7)

While they will come to the fore in the romances both as a threat and a promise, the political entailments of that undoing of the founding relation between art and law are already intimated by *Twelfth Night*'s most extravagantly self-referential moment. The text of Maria's forged letter runs:

> I may command where I adore,
> But silence, like a Lucrece knife,

With bloodless stroke my heart doth gore.
M.O.A.I. doth sway my life.

(*Twelfth Night*, 2.3.94–97)

The violence of the "Lucrece knife"—the weapon with which the chaste noblewoman had committed suicide after her rape by Tarquin—had already made its appearance with Malvolio's forcible opening of the chastely sealed letter: "By your leave, wax. Soft—and the impressure her Lucrece, with which she uses to seal. 'Tis my lady!" (2.5.83–85). In fact, Malvolio had anticipated the incision even earlier. Taking up the closed letter, and betraying his own obscene preoccupations, he declaims, "By my life, this is my lady's hand. These be her very c's, her u's and her t's, and thus makes she her great P's" (77–79).

Such reiterations bear out the cut's instituting status. Insofar as the division that forcibly unseals meaning is indistinguishable from the meaning it would expose, reading enacts what it would comprehend: Thus the inscribing redounds. But what's important in this instance is the temporality of the effect. Marking a division in space—Malvolio's reading is situated fully outside the text (none of the letters he alludes to actually appear in the superscript) even as it already foresees its content—the cut equally displaces itself in time, appearing from the outset already ahead and behind itself. "And then 'I' comes behind," Malvolio declares. An act that undoes itself as an act, a "silent" wound outside the phenomenal register altogether, the self-rivening "event" that constitutes the text as a coherent speculative domain at the same time unbinds that self-reflective aesthetic logic, reducing every identity to the sheerly material trace of a missed or anticipated occurrence. "If thou hast her not i'th' end, call me cut," revolving Toby declares, a line that could have been voiced by the no less precipitate, belated, and queer caesarian maid, as well as by the theatergoer who would have and will be had "i'th' end" (2.3.170–71).

Of a piece with the radically arbitrary and nonhuman positing entailed in constituting a relation to relationality as such, such disturbances occur at the point where art and law converge and where the polis itself is at issue. Shakespeare's audience would have recognized in chaste Lucrece's sacrificial suicide the act that founds the Roman Republic. A self-violation in the name of intact purity, that act conveys the indissociability of law and transgression even as it figures in advance the redounding violence that seeks to apprehend it. Precisely at its lightest and most self-conscious, Shakespearean comedy is implicated in that contradictory founding violence. The knowing unknowingness—the claim to compass one's own epistemological limits—of aesthetic ideology is staked on a compulsion less stable than the forms of exclusion—the "thing[s] of darkness"—such tolerant knowingness acknowledges, a force that in this instance assumes a gendered form. In that regard, Shakespearean comedy has something to say about the surprising persistence of forms of sexual and racial violence within the aesthetic state, which is to say, the liberal personal and political dispensation that confirms itself in the acknowledgment of its inherently constructed condition.[12]

2

Romance extends comedy's "thither and back again" beyond death's limit, law's founding bar now explicitly aligned with that seemingly absolute mortal boundary, and in the process deepens the psychic and political stakes merely intimated in the comedies, even as in the exorbitance of its aesthetic claims it extends those stakes more overtly to questions of world creation.

Though such genealogical accounts are unfashionable, with critical focus turning in recent years to the importance of broader shifts in popular taste toward Italian tragicomedy as well as popular prose romance,[13] it is not wrong to see in Shakespeare's late romances, with their various miraculous returns from the deeps and from the dead, something like a sublimating passage beyond the fatal limits of the tragedies, desire's bar now translated—or failing to translate—into the condition of a certain reconciliatory transcendence.

The generative impossibility of such a sublating process lies at the heart of the genre's contradictory formal traits. None of Shakespeare's dramas are as tightly controlled and reflexively aestheticizing as *The Winter's Tale* and *The Tempest*, the latter explicitly observing the classical unities. And none are more dispersed and tonally unstable, *The Tempest* with its ambient voices and expansive web of specular correspondences in which characters eerily anticipate and reiterate themselves and one another. Implying a transformed understanding of character and figuration, such traits index the romances' status as radically aesthetic formations. According to the logic of such autonomous formation, the work constitutes itself as a work by virtue of reflexively comprehending its own bounds and limits, whether that's a matter of the formal reflexive mediation of a choral figure—Gower, "Time"—or the incorporated end of a death that will have been no death at all. And yet, because it is indeed instituted in that autopoietic self-exorbitancy—because it emerges out of it, just as Time, "self-born," constitutes itself in the "wide gap" of the play's time—the work exceeds and disperses itself in the very movement of its consolidation (*The Winter's Tale*, 4.1.7,8). The opaque beyond against which romance forms itself—the unseen storm that precipitates *The Tempest*—inevitably returns within as the divided condition of all-encompassing aesthetic consciousness itself: "And now I pray you, sir, / For still 'tis beating in my mind, your reason / For raising this sea storm?" Miranda says at play's start; "A turn or two I'll walk / To still my beating mind," Prospero says toward the close, breaking the spell of his masque even as he reconstitutes that spell in the ravishing image of its dissolution (1.2.175–77, 4.1.163–64).

The constitutive limit—law's bar—that had been a diegetic element of the comedies is now absolutized as the condition of the work itself: thus the striking convergence of formal and thematic concerns in the romances, their persistent feel of allegory. The autonomy of the work is mirrored in the plays' preoccupation with the protagonists' autonomous or autogenetic beginnings: the sovereign magus's world-creative potencies, Posthumus's ex nihilo birth from parents "deceased / As he was born" (*Cymbeline*, 1.1.39–40), or sea-born Marina, whose parentage "did end / The minute [she] began" (*Pericles*, 5.1.200–1). At the same time, the splitting that is the condition of that self-incorporative wholeness informs the texture of the work at every level, from the temporal "gap" of choral interventions and narrative leaps, to the lost and refound child's birth "between two shores," to the restraining, nonpossessive distance enjoined upon Leontes before the returning image of his wife, figure for the proximity and distance of aesthetic apprehension itself.

According to the contradictions out of which it emerges, the autonomous work—the work that would compass its own grounds—is also the work that most condemns itself to uncontained reiteration. Like Oedipus, Pericles's saving flight from incestuous prospects merely blinds him to the way he repeats what he flees: "Oh come hither," Pericles declares to his newly recovered daughter, "Thou that begett'st him that did thee beget" (5.1.185). Posthumus resolvingly discovers the wife he'd presumed he had killed precisely at the point he repudiates her anew: "You ne'er killed Imogen till now," Pisanio exclaims, after Posthumus strikes the disguised figure and she falls (5.5.231). Such a coincidence of cure and ailment arises because, according to the autopoietic logic of constitutive difference, recognition and refusal are one and the same. "When I did push thee back— / Which was when I perceived thee," declares Pericles to his daughter (5.1.117–18). In *The Tempest*, the moment of Prospero's most outright loss of potency—the forgetfulness he suffers during the marriage masque he stages that prompts his beating mind to thoughts of death—is also the moment of that power's fulfillment in the form of the absolutely inclusive work, the work that absorbs its own creator. And with that coincidence, mastering self-completion coincides with limitlessly reiterative inscription, Prospero's self-captivation repeating the blind absorption in his studies that led to Prospero's original usurpation, setting the revenge plot in motion.

Cause of its unraveling, such contradictions are also the seed from which narrative emerges. The inapprehensible threat of incest that opens *Pericles*—to speak the riddle's solution is merely to renew the evil, Pericles declares—equates the sovereign dream of absolute, time-defeating self-completion with a dissolution of the distinction between self and other that would undo origination altogether, including narrative origination. The play's abrupt advent—the flight from the mere thought of incest that sets in motion the drama's episodic adventures—amounts to the condition for narrative as such. "Running from both tyranny and his own generic bewilderment, Pericles must in a sense be built from nothing," Simon Palfrey observes.[14] Such precipitancy is the condition of sovereign agency itself in the plays. *The Winter's Tale*'s drama emerges as if out of the blue, with Leontes, like an Iago-less Othello, instituting the scene that undoes him. "I am angling now, / Though you perceive me not how I give line. / Go to, go to!" he says as he sends Hermione and his rival off to the garden: "[It is] as he had. . . been an instrument / To vice [them] to't," Camillo remarks (1.2.179–81; 1.2.413–14). The self-division that characterizes that beginning is indissociable from the divided condition of the king's intrinsically spectatorial experience of it. "Go play, boy, play.—Thy mother plays, and I / Play too, but so disgraced a part, whose issue / will hiss me to my grave," Leontes declares at the start, as if already speaking from beyond play's end, enacting the self-fulfilling and self-consuming theater of sovereign power in its purest form (1.2.186–88).

Originary and generative, such vicarity becomes an ontological condition in the romances, bearing out aestheticization's inextricable relation to phenomenal experience as such. The precipitating "event" in *Cymbeline* occurs when Giacomo invades sleeping Imogen's chamber not to commit rape but to gather the evidence to produce the tale of such a crime. And what he encounters as proof is a fictive scene that already anticipates him: the tapestries displaying the scene of Tarquin's assault, the book Imogen had been reading with page turned down at the scene of Philomela's rape. According to the causeless causation of the work's endlessly substitutive universe—adversarial Cloten a stand-in for Posthumus, Posthumus for Cloten—every event transpires as if too late and too soon.[15]

And yet it is just that untimeliness that gives the drama its preternatural immediacy, exposing what has always been the secret grounds of theater's affective claims. With Giacomo's reflexive and violating intrusion—his invasive reading of the signs of his deed—the "original" sin of the play coincides with the audience's own pruriently intrusive and self-violating absorption. With that convergence, the elliptical condition of the work as aesthetic formation, its constitution in the movement beyond itself and as an effect of our own interpretive intervention, proves the condition of sensual experience as such, the movement by which sense comes to—"touches"—itself in exceeding itself.[16] "Love's counselor should fill the bores of hearing / To th' smothering of the sense," Imogen declares, evoking the ecstatic "beyond beyond" of her anticipated joy, just as the sensory immediacy of *The Tempest*'s phenomenal world is never more felt than in Caliban's transporting account of a dream of bereaved waking and of waking to dream again (3.2.56, 57–58).[17] That the most artificed of Shakespeare's works should also be the most concerned with the immersive "sounds and sweet airs" of the human and inhuman sensorium merely bears out the connection between aestheticization and the possibility of phenomenal experience itself, a phenomenality that institutes itself precisely in its movement beyond human limits.

Penetrating the intimate dimensions of sense apprehension, aestheticization in the self-exceedingly incorporative form it assumes in the late romances at the same time fulfills itself in the work's "outward" extension into the world beyond it, including a political domain more explicitly engaged here than in the comedies. In *The Winter's Tale*, the work's aesthetic form is indissociable from matters of law and sovereign authority; in *Cymbeline*, aestheticization is bound up with the problematic foundation of the nation. The tension in that latter play between Britain's claim to autonomy and its claim to a rightful place within the Roman imperium can be understood in terms of the era's desire to recover national continuity even while exorcising the "primitive savagery [it] wished to declare obsolete."[18] And yet such tensions equally reflect the contradictions of the modern nation-state as such, the problem of the nation as a purportedly organic form nevertheless constituted within the purely relational framework of the modern geopolitical domain, where, absent transcendental anchoring, every political entity will be defined vis-à-vis its finite boundaries.[19] In such a dispensation, every nation must aspire to be "of . . . the world's volume . . . but not in't," as Imogen says of her homeland—within and beyond the system of relations that defines it (3.4.137–38). That at once inviolate and bordering condition underlies the ease and frequency with which the protagonist switches sides in the conflict, as well as the curiously talismanic status of "Milford Haven," the obsessive invocation of that place name bearing on the liminal character of the port as the point of crossover between times—Rome's invasion in ancient prehistory and Henry Tudor's returning to claim the throne in the modern era—between internal and external broils, even between psyche and world: Imogen calls out the name as she lingers between waking and sleep, the place that lies "without me as within me"—the psychic space where theater itself transpires (4.2.306).[20]

In part, such contradictions are resolved thematically. The notable fact that the most full-throated celebration of Britain's self-sufficient purity is voiced by the evil stepmother queen can be understood in terms of the plot's structural logic: By casting out that figure as a figure of emasculating gender dependency, the play sublimates—negates and preserves—autonomy now in the form of self-sufficient male bonds that cut across the boundaries between Britain and Rome; with that, Cymbeline will declare himself "mother to the birth" of all his ruling fold. Ultimately, though, the resolution of the nation form is staked precisely on the unreadability of the contradiction that determines it. When called on in advance of the final battle to interpret his dream, the soothsayer finds in the vision of the Roman eagle's "vanish[ing] in the sunbeams" a portent of Rome's victory. After the Roman loss, he rereads the dream: "Soaring aloft" "from south to west," the eagle "lessened herself, and in the beams o'th'sun / So vanished," thus "foreshow[ing]" that "Th'imperial Caesar should again unite / His favor with the radiant Cymbeline, / Which shines here in the west" (5.5.469–74). Irresolvable division and loss are translated into perspectival diminishment, the consolidated nation taking on its true and necessary form as vanishing point, that which is always yet to come.[21] With that, the nation itself proves an effect of the process of aesthetic phenomenalization, the mechanism by which division is internalized and thus recuperated as subjective inwardness. Had I seen him depart, Imogen says of the ever-parting Posthumous, "I would have broke mine eyestrings, cracked them, but / To look upon him, till the diminution / Of space had pointed him sharp as my needle; / Nay, followed him till he had melted from / The smallness of a gnat to air, and then / Have turned mine eye and wept" (1.3.17–23).[22]

The violence of Imogen's figure of eye extrusion betrays the fragility of such a consolidating mechanism, suggesting it may be impossible to cognize such horizon effects without falling subject to their divisive force. As it was in *The Merchant of Venice*, the limits of autopoiesis in the late plays are most directly manifest in their relation to their defining counterterm: law. In the earlier comedy, the play's powers of aesthetic suspension—of "tarrying"—come into play in relation to managing the force of law as a thematic concern within the play. In the later plays, the relation between art and law is explicitly staged as the condition of the work itself as aesthetic form, a fact that becomes especially apparent at the point where the grounds of theater's powers of phenomenal conjuring are most self-consciously realized: the scene of Hermione's return as a living statue at the close of *The Winter's Tale*.

That scene of artful return answers to the play's earlier evocation of law's supervening force in the oracle's annihilating pronouncement. The structure is not a simple opposition, the finitude of law as against the free play of fiction. Law, too, is a definitionally indeterminate condition. The uncanny seamlessness with which we move from Leontes's refusal of the oracle's judgment to the realization of that prophecy in Mamillius's death reveals the force of a law that, as in tragedy's ensnarements, is constituted precisely in its refusal, thus realizing its inescapable power to annul subjective agency and inscribe itself within: "But of all, the burst / And the ear-deafening voice o'th'oracle, / Kin to Jove's thunder, so surprised my sense / That I was nothing" (*The Winter's Tale*, 3.1.8–11). More explicitly than in *Merchant*, law enacts its own autopoietic condition.

Art doesn't banish such reiterative effects—it translates them into its own knowing condition, the play thus staging comedy's autopoietic redoubling of tragic autopoiesis within its own unfolding narrative form, even as it seeks to fulfill aestheticization's impossible ambition of compassing its own grounds by incorporating law's absolute limit. The scene of Hermione's elusive and artful return corrects the redounding violence of the effort to force a perfect resolution suggested by the prior, reported scene of Perdita's recovery, where the demand for absolute proofs and clinchingly specular recognitions amounts to a self-rending—"they seemed almost, with staring on one another, to tear the cases of their eyes"—and where the force of the desire to fend off loss threatens a no less tyrannical repetition of the loss—Paulina "lifted the Princess from the earth, and so locks her in embracing as if she would pin her to her heart, that she might no more be in danger of losing" (5.2.11–12, 71–73).

Hermione's return ameliorates that redundantly possessive violence by consciously sustaining the gap opened by its own status as aesthetic representation. Through the reflexive movement by which it both exceeds and comprehends itself as art, the play at once grasps and reiterates the division that leaves Hermione, as it leaves us, suspended between loss and return, death and life, work and world, absorption and release—the space of aesthetic consciousness as such. "Let be, let be. / Would I were dead but that methinks … it breathed," Leontes declares (5.3.61–64). In other words, the play constitutes all in the knowing and absorbed condition we have inhabited all along as theatergoers. On stage and beyond, all are lookers—"I am content to look on," says Leontes; "So long could I / Stand by, a looker on," says Perdita—and when Hermione herself, the object of every demanding gaze, finally speaks, she only gives back her interlocutors' own question in inverted form: "Tell me, mine own / Where hast thou been preserved?" she says to her daughter: It's a line addressed no less to us at the point of our own enigmatic coming-to at play's end (5.3.92, 84–85). Each witness to their own witnessing, all are fragilely held within an aesthetic community of sheer vicarity, bound with one another insofar as each is absorptively divided from themself.

The possibility of such an apparently inclusive space of the knowingly unknowing hangs ultimately on art's defining relation to law. In *The Winter's Tale*, killing law and meliorative art are articulated via their symmetrical and mutually defining opposition across the two halves of the play, even as the play claims that very division as a function of its own aesthetic design. Through such an incorporated divide, art constitutes itself in its delimitably all-compassing aesthetic form, even as the inhuman, desubjectivizing force of law is reflectively phenomenalized and thus made humanly available—by virtue of art, autopoiesis is apparently made available to consciousness. "If this be magic," Leontes says of the final transformation, "let it be an art / Lawful as eating" (5.3.110–11). As art and law converge, the very distinction between law and vital experience approaches a blessed dissolve.

And yet that very convergence poses a threat more dire than law's tragically mortal hold and perhaps a promise more radical than aesthetic communion, a threat and a promise felt in the mingling of art, law, and eating. Autolychus, the self-making trickster, utters his verses "as he had eaten ballads" (4.4.185–86). At the point where it absorbs its own boundaries, consumed art threatens to consume all. The chaos of an heirless sovereign will descend on the kingdom and "devour / Incertain lookers-on" (5.1.28–29). "A grace it had, devouring," Prospero says of the conjured banquet that has held his conscience-stricken victims fast, devouring even in its being devoured (*Tempest*, 4.1.85). What is consumed thus is nothing, death—"What, must our mouths be cold?" the storm-tossed mariner declares—and everything: reflexive consciousness as such. "They devour their reason," Prospero says of the wide-eyed and unseeing audience he gathers at play's end (5.1.155).

Art devours at the point where, under the pressure of its totalizing ambition, any governable relation between law and aesthetics dissolves. In comedy, the inapprehensibility of the limit that defines fiction as fiction is narrativized—we will always miss the magician in the forest but he was there for all to see. In the romances, where that constitutive limit coincides with the ontological boundary of life, the "wide gap" of that missed ground undoes narrative altogether; we can't know whether Hermione is returning or has always been present elsewhere. "It is required / You do awake your faith," Paulina says (*The Winter's Tale*, 5.3.94–95). That caesura at the generative heart of the work remains a matter of faith because it is what opens the very distinction between fictive and real, and with it, the conditional condition of phenomenal experience itself. But precisely because it is constitutive, we can never be certain whether such a positing leap of faith is ours to make or whether like law insofar as we are subjects to it, it befalls us. Conscience itself, that which internalizes and thus humanizes the fatal dictates of law, translating law into spirit, redounds like something poisonously extrinsic and alien: "Do not infest your mind with beating on / The strangeness of this business," Prospero says to his spell-bound adversaries (*The Tempest*, 5.1.249–50).

Such violence isn't limited to the play's miscreants—witness Prospero's final call for our "pierc[ing]" prayer. As we noted in the Introduction, when in the epilogue Prospero requires our applause to draw his project to a close—the "gentle breath" of clapping that will fill his departing sails and "release [him] from [his] bands"—we may well be tempted to a more fiercely punctual response, an aggressively breaking the spell in order to reclaim ourselves from its hold. To do so is, we noted, merely to return to the beating tempest with which the play began, and thus to renew the work's hold. The solution, what unbinds that redounding economy, and with it the recoils of revenge from which the play has sought release, is the gentler accommodations of aesthetic consciousness: to hold onto the liminal space in which we hear the sound of our own clapping—what we choose and what echoingly returns, bringing us to as much as it brings the play to a close—as what comes to us from elsewhere, as the familiar and strange sound of a beating in mind that is the enigmatic source of aesthetic causation. To hold open that inward and alien space—the space of theater—is to reclaim loss as a higher knowing unknowing.

And yet, precisely in engaging aesthetics at its rarified limit—as subjectivity's cause—the romances court what remains inassimilable to even such a self-abrogating claim to cognitive mastery: the divisively material force that troubles even as it enables such apprehension. Confounding the punitive force of law and the pulse of life itself, the beating mind ultimately suggests a violence that can't be separated from the generative power of art. "What fine chisel / Could ever yet cut breath?" Leontes says of the living statue, a sublime figure for the sheerest limits of aesthetic refinement, art's accommodation to spirit, and a terrifying figure suggestive of the material cut of a Lucrece knife within breath itself, and thus the silent wound from which all the echoing apotropaic violence in and beyond the play devolves (5.3.78–89).

That exorbitancy at the heart of aesthetic formation bears on the question of its historical status. As Cymbeline suggests, aesthetic phenomenalization plays a direct role in the ideological formation of the nation, the knowing unknowingness of aesthetic consciousness reflecting the necessarily suspended condition of the nation-state, a formation always still to come. As we have noted, the divided totality and limitable limitlessness of aesthetic consciousness are consonant with the process through which the early modern social domain, untethered from certain transcendental grounding, articulates itself around its own founding divisions, in relation to what Lefort calls the "enigma" of its "external/internal articulation" as a self-constituting sphere.[23] But such claims for the causes or aims of aestheticization are necessarily equivocal.

Bound up as it is with the grounds of phenomenalization, aesthetic formation is indissociable from the process by which history is given a narratable form, translated into a narrative of causes. At the same time, precisely at the point where its compassing ambitions are most fully realized, the process of aestheticization registers history otherwise, as a contingent force, a non-narratable "cut," that unhinges sense experience itself.

Articulating the autopoietic logic of aesthetic formation at its foundation, Shakespeare's drama exposes its defining contradictions, in the process intimating the potential futures harbored within our own fragile and contradictory political-aesthetic dispensation. The foundering of the opposition between law and aestheticization opens, on the one hand, the prospect of a radically open and untotalizable relationality, where, in place of a judicial economy based on a sovereign assignation of guilt and innocence and thus vulnerable to the redounding instabilities of such founding prohibitions, a limitless fictive law would presume exposure and vulnerability as the inaugurating condition of every subject, the mark of the finitude and vicarity that determine each as an irreducibly relational being.[24] Such exposure isn't a matter of being open to the world; it is what constitutes world as being-with and being elsewhere.[25] On the other hand, an art as lawful as eating, one in which the distinction between law and aestheticization approaches a dissolve, equally suggests how proximate the knowing unknowingness of rarified aesthetic consciousness can be to its totalitarian inverse in the disastrous equivalence between law and corporeal embodiment, law understood, for instance, as the carnal voice of the ruler and thus as a devouring annulment of the scene beyond our own that opens the very space of the world.[26]

5

Hurt Feelings

Affect, World, and Time in As You Like It

If the autopoietics of tragedy suggests the inhuman character of pathos, comedy, with its more thoroughgoing engagement with the conditions of social relationality, clarifies the broader relation between affect and world creation. Recent work in the field has cast that relation via a notion of affect as a form of "embodied cognition," a characterization that is taken to suggest affect's ecological promise. Here, focusing on Shakespeare's *As You Like It* and taking my cue especially from the work of Jean-Luc Nancy, I'll argue that affect becomes affect precisely insofar as it exceeds any such embodied or immanent condition. Even as it implies a heterogeneity that insistently extends the subject beyond itself, affect's constitutive divisions and displacements baffle any attempt to understand that relation between self and a world beyond as a matter of sympathetic correspondence, the kind of mimetic trope that informs many ecologically oriented accounts. Rather than revealing the subject's correspondence with or embeddedness in an already given world, affect understood as an autopoietic phenomenon, or one that touches the limits of autopoiesis, bears on the prior and more fundamental constitution of those very categories of self and world: Rather than reflecting the world, affect concerns the opening of selves and worlds.

I'll focus on a punctual moment in Shakespearean comedy: Rosalind's faint in *As You Like It*. A conversion scene involving the metamorphosis of selves, the scene simultaneously involves the re-creation of the referential world within which it unfolds even as it extends its contagiously transformative effects to the audience beyond—in that regard it anticipates the world-creative dimensions of the romances in their most "ontologically" expansive form. Suggesting the grounds of theater's affective hold, the scene at the same time reveals in its very theatricality a vicarity at the core of affect: Paradoxically, theater precisely as theater draws us toward the grounds of affective life as such. The potency of such an affective moment—its power to inscribe all in and beyond the stage—is inseparable from the elliptical character of its unfolding. Reflexively constituting the world to which it at the same time responds, the temporally riven response the scene provokes bears out affect's distinctive relation to what Heidegger terms our "thrown" condition, the fact of self and world alike being always already and illimitably entangled. Implying a temporal division and displacement inhabiting the moment of response, the scene suggests that affect isn't just a historically conditioned phenomenon: It is indissociable from the fact of historicity itself.

1

The recent turn to affect in cultural and literary studies has been inspired in part by what it promises for an expanded understanding of human subjectivity in its relation to the world, one that thinks beyond the autonomous, ratiocinative being of our presumed Cartesian inheritance—thus affect's special pertinence for early modern studies. Conceived as a matter of bodily intensities rather than cognition, affect is seen to orient us "beyond the boundaries of the singular subject," becoming, in Steven Mullaney's formulation, a matter of "embodied social thought."[1] As definitionally relational, affect reconceives the human subject as necessarily inscribed within the network that defines it, whether that domain is imagined as a socius or a larger ecological field. "Affect is persistent proof of the body's never less than ongoing immersion in and among the world's obstinacies and rhythms."[2]

The connection between embodiment and an environmental conception would seem to make intuitive sense insofar as material embodiment suggests the promise of affect's extension beyond the dictates of the autonomously conceived Cartesian cogito. Evelyn Tribble and John Sutton speak of the "embodied and extended model of cognition" implied by relationally conceived knowledge, and Amanda Bailey and Mario DiGangi describe affect in their introduction to a recent collection on the topic as the "embodied manifestation of the interface between individuals and their environments."[3] And yet, as I noted in the Introduction, that association between "embodiment"—the term that echoes through many such accounts—and relational sociality remains problematic.[4] For Bailey and DiGangi, the latter is critical in order to avoid affect study's regression to "essentialist ideolog[y]": Affect encourages us to "consider ontology as a relational proposition," a critical expedient lest affect be no more heteronomous than the atomized subject of reason.[5] And yet, if we take the idea of relational constitution seriously, then affective experience becomes purely a function of the mediatory division or "interface" that produces it and thus no longer "irreducibly bodily" at all.[6]

The problem of relation recurs in recent efforts to claim for theater a phenomenological recovery of "the wholeness of being that each human speaker knew before language."[7] For Bruce R. Smith, that unmediated condition, with its promise for a "thoroughly physical, thoroughly grounded" version of theatrical experience, is entirely consistent with a recognition of the subject as a relational being. But if the subject is a function of "relational ways of knowing," as Smith suggests, if it comprehends itself via such relationality, it becomes hard to discern in what sense it amounts to a fully unmediated being.[8]

The tensions implicit in the very notion of an "embodied . . . interface" tend to be resolved through reference to the familiar emotion-affect distinction, that is, the difference between emotion understood as a function of the representational and thus mediatory logic of a consciously or unconsciously intending subject and affect as self-immanent presubjective life force. And yet the distinction may itself prove difficult to sustain. As Rei Terada shows, even in its classic philosophical acceptation, emotion by definition exceeds the ambit of a cognizing subjectivity: What there is of mediation in that context is not a matter of the cogito.[9] At the same time, constitutive difference tends to insinuate itself even into accounts most committed to a vitalistic and antirepresentational understanding of affect.

Giulio J. Pertile's claims for an autonomous affective "life stream" coursing beneath consciousness borrows from phenomenological descriptions of the autoaffective grounds of cognition. "Subjective experience feels itself before it relates to its object," and it does so, Pertile argues, by virtue of what Michel Henry describes as "a power in which it is auto-affected so that it feels its seeing and feels itself seeing."[10] "Henry's version of the Cartesian cogito perceives itself because it is, always already, fully inside itself, because it perpetually encounters itself through an intimate self-touching that bypasses every form of reflection and representation"—a monistic affective subject fully independent of the representational logic of cognition.[11] And yet, insofar as it involves the reflex of self-touching, insofar as touch begins as touching itself touching, such autoaffective origination necessarily implies a division and displacement that troubles the distinction between intimacy and distance, as well as the claim that affect occurs "inside" anything at all. If, as Henry claims, "it is in the experience of a radically immanent subject that life arrives to itself and seizes its being" via the autoaffective touch—if it is inaugurated in that arrival—then affect will have always come to itself from elsewhere and will inevitably be marked by the sign of its secondarity.[12] Far from implying a monistic and substantive life stream or the promise of "lived forms of meaning," affect is from the outset a function of the passage beyond itself through which it will have been called forth, a function, that is, of its own finitude and limit.

The enigma of self-relation persists even for phenomenological accounts rigorously attentive to the problem of mediation. Arguing against Bernhard Stiegler's claims for the determining significance of cinema in the affective structuring of modern time-consciousness, Mark Hansen asserts that, rather than assuming a constitutive role, technicity merely intervenes within or "contaminates," even as it expands, affect as an already embodied "living presence." He finds support for that claim in Edmund Husserl's account of the double nature of consciousness's relation to the temporal flow that constitutes it as consciousness. "'Every experience,' Husserl explains, 'is "consciousness" [*Bewusstsein*] and consciousness is always consciousness-of,' but every experience is also 'itself experienced [*selbst erlebt*] and to that extent also intended [*bewusst*]. This being intended [*Bewusst-sein*] is consciousness of the experience [*Erlebnis*].'"[13] Consciousness may be a function of the fact of temporal deferral and flow, but it must also entail an awareness of that perception as perception—it must be conscious of itself experiencing. In experience thus objectlessly intending itself Husserl locates the primacy of the "now," of the "substance of the living present," in the determination of consciousness.[14]

And yet, it's not hard to sense the recursive regress in such an account. If intention means the mechanism through which experience is "itself experienced," then such experience already harbors within itself as a function of that reflex the difference and deferral that erodes it as presence, that makes the perception of temporal flux itself a function of such flux. The strain can be felt when, having here spoken of the primacy of the presently intended perception of flux, Hansen will later argue that perception itself is a function of temporal deferral, anticipatory "protention" now becoming what "intends the new prior to any impression or perceptual present."[15] If intention protends, it anticipates itself, a condition that would seem to undo intention as a vectored, self-present structure.

Insofar as affect's autoaffective formation at once opens and troubles the opposition between subject and object, intimacy and distance, it suggests that affect might indeed have a privileged role in reimagining an expanded conception of being. But it does so in a way quite distinct from claims for "sympathetic correspondence" that inform many accounts of the ecological promise of affective embodiment, with the implicitly mimetic logic such correspondence presumes. Gail Kern Paster initiates her reading of theater as an expansive "affective-cognitive ecology" by citing Jean-Luc Nancy's account of the constitutive force of relationality—the "among-being of all beings"—reading it as an argument for the "inter-subjective" status of meaning.[16] By "intersubjective"—a recurrent term in many eco-affective accounts—Paster means the forms of inference and "presupposal" by which subjects convey their "singular" "innate dispositions and personal habits," producing a "shared social intelligence of each other's habits and casts of minds."[17]

And yet, to read Nancy as an advocate for intersubjectivity is to misconstrue what he means by the "among-being" of being. Insofar as it concerns a relationality and condition of responding in advance of cognition, "being-with" for Nancy implies the indissociability of being—the very being of the world—and affect. Such being-with is not, however, a matter of sharing already given dispositions, or of "intensities that pass body to body," or even of mutually determined meanings or dispositions.[18] Rather, being is constituted in the first instance by virtue of the mediatory "with" as such, a condition that conjoins being only insofar as it at the same time divides and endlessly displaces it in relation to itself, resisting any form of collective totalization whether as a social group, an "affective-cognitive ecology," or an accretion of discrete empirical elements, however expansively conceived.[19] Implying neither individuation nor totality, "singularity" for Nancy entails the originary limit by which affective being comes to or disposes itself—assumes its proper immanence—solely by virtue of "revealing itself beside, always beside."[20] Affective experience, then, isn't a matter of correspondence, whether read at the level of an "animate, interconnected universe" or of the "sympathetic correspondences" taken to represent the space of the stage.[21] Instead, affect is an experience of incommensurability. "Emotion arises because of a lack of fit between mind and world and between the mind and itself," a noncoincidence that at the same time opens the relation between self and world and thus amounts to the condition for any form of rapport.[22]

Such inadequation isn't a failure—it's not an inadequacy *to* anything. It is the reflexive ellipsis by which being as such appears at all—the ever-repeated autopoietic origin of world, in Nancy's account. "At the point where we would expect 'something,' a substance or a procedure, a principle or an end . . . there is nothing but . . . the turn of the other access, which conceals itself in the very gesture wherein it offers itself to us," a nonencounter that is the condition for any encounter at all.[23] References to the "affective turn" in critical studies are rigorously apt, for affect is nothing other than such a turning, a condition that makes affect at once the object and never quite the object of its own theorization. The "something" that does and doesn't materialize by virtue of such an elliptical turning should be understood as any substantified variant of affect, whether "the intersubjective stuff of feeling," the "visceral forces beneath, alongside, or generally *other than* conscious knowing" or even as a "*beside-ness*" capable of "accumulative" presence or persistence.[24]

The relation between affect and the always incomplete process of materialization makes apparent the derived or fictive nature of affective "embodiment."[25] To recognize that derivativeness, to refuse a conception of affect as organically embodied or self-immanent, is to acknowledge the fact of its thoroughgoing historicity. One might be tempted to read the relation between "sympathetic correspondence" and Nancy's "being-with" as a difference between historical epistemology—the early modern regime of resemblances as Michel Foucault, for one, articulates it—and a universalist or presentist analysis. But Nancy isn't presentist or universalizing. The division and displacement he ascribes to affective experience is the mark of its inescapable historicity, and the relation between mimesis and difference implied by that fact of historicity is more complex than a simple opposition. In *The Reformation of Emotions*, Steven Mullaney is scrupulously attentive to the contingent and overdetermined character of affect as a historical phenomenon, arguing explicitly against "affective mimesis"—the idea that affective expressions provide unmediated access to how people truly felt.[26] It's worth noting, then, the way the mimetic impulse nevertheless returns in his account when he construes this very indeterminacy and groundlessness as an embodiment of the spirit of the age, an expression of the trauma of the Reformation moment. Such slippage isn't an error, correctible by reference to some purer account. It is the movement of affect itself, the turn by which affect at once reflectively realizes and displaces itself, producing the effect of phenomenalization through which anything like a "felt history" or "felt relations with the past" is constituted.[27]

The world-creative dimensions of Nancy's account of being as "being-with" bring us back to the question of the ecological entailments of affect, suggesting affect's potential for rethinking what we mean by world. It is striking how frequently accounts committed to undoing the givenness of a self-enclosed Cartesian subject affirm in their very reference to the "embedded" nature of that contingent being the givenness of the world as an enclosing domain, thus leaving intact the doublet by which self and world have always reinforced each other. The prior assumption of the space of the world can be explicit in some phenomenologically oriented accounts; however relationally conceived consciousness is in Smith's analysis, the subject remains an expressive being oriented within the terms of an already determined inside and outside, a matter of the "out-flowing" or "externalization of passions."[28]

But the status of world remains an issue even in accounts more committed to a conception of the subject as the function of relational networks. The sheer range and varying registers of the potential elements in the "cognitive ecology" forwarded by Evelyn Tribble and John Sutton—"neural and psychological mechanisms," "bodily and gestural norms," "parts, plots and playbooks," "sound and lighting," social and economic systems, "directorial control, clowning, visuality and improvisation"—make evident how much any possibility of meaning in such a relational system, insofar as it is genuinely relationally determined and not just accretive, will be a function of the interpretive act involved in determining its bounds, in arbitrarily deciding what's included and excluded, for it is solely by virtue of such a limit that the system becomes a meaningful system at all.[29] At the same time, it is by virtue of that forced limit that the subject instates itself as cognizing subject, as the being who comprehends the objectified world as a world. Thus, the paradoxical complicity between "immanence" understood as an autonomous and substantive domain—a "life stream" or an "integrative" accumulation of empirical elements, or even, I would argue, a Deleuzian "plane of immanence"—and transcendence.[30] "Why are we so keen to distinguish the human self from the field?" Jane Bennett asks.[31] Perhaps because to imagine we can see ourselves as part of the field and still comprehend it as a field, or to imagine we can descriptively grasp ourselves at once "by and as" affective experience, is to fulfill the transcendent fantasy of self-inclusive comprehension in its classic form.[32] Immanence rigorously conceived, which is to say immanence as an expression of the fact of illimitable relationality, necessarily implies the divisive reflex through which that field assumes and constantly fails to assume phenomenal form, becomes and fails to become a field at all. In that sense, immanence means not vital life but finitude and appropriation—no end of ending.

To discern what such an understanding of affective "being-with" would mean in social and ecological terms, we need to recognize the double methodological condition it imposes. On the one hand, we cannot simply step outside consciousness in order to objectively describe the network of relations within which it is inscribed.[33] To do so would imply a transcendence that refuses the very condition of relational constitution: "Being as being-with might no longer be able to say itself in the third person, as in 'it is' or 'there is.' Because there would no longer be a point of view that is exterior to being-together."[34] At the same time, insofar as it is an immanent function of that differential web, being is never simply "within" anything, least of all itself. As what is from the outset always within and beyond what determines it, affective being doesn't become social or ecological by extending itself toward the world, by expanding to include human or inhuman others. It has already made such a passage, intersecting the other precisely at the point where it will have insistently failed to coincide with itself, a missed encounter and temporal dehiscence from which self and world alike at each turn emerge.[35]

2

In order to see what such an autopoietic conception of affect might look like in practice and how it bears on matters of time and world, I want to turn to a scene from Shakespeare's comedies, one that has something to tell us about the grounds of theater's affective hold. Obviously, approaching affective relation through a domain—the literary—that is by definition mediated and vicarious would seem to beg the question. Then again, precisely insofar as it opens a supplementary aesthetic space between self and world, one belonging neither quite to the one nor to the other, the literary as literary may in fact have a distinctive power to figure the condition of "being-with." Literature, that purely ancillary diversion that takes us from the world and ourselves, at the same time draws us closer to the grounds of self and world, I'll suggest. The fact of theater's status as "embodied" fiction merely heightens its disquieting capacity to expose the illusory or contrived character of phenomenal experience as such.[36] The moment I'll engage—the scene in *As You Like It* in which Rosalind, hearing the tale of Orlando's encounter with a lioness in the forest and being shown the handkerchief bloodied by his wounds, faints away—bears out Pertile's insight that the nature of affect may be most revealed at the point of its failure, though I won't argue as he does that that is because insensateness reveals a vital affective current beneath the threshold of conscious experience. Instead, such a lapsus defines affect as such and in doing so conveys affective being not as an eternal substrate but in its irreducibly historical character.[37]

This is literature, of course, but Rosalind's swoon might nevertheless suggest an argument for the embodied character of affective experience. After the array of fictions she has set in motion—Rosalind has been playing the part of Ganymede playing the part of Rosalind in a love game with Orlando, with whom she is in fact smitten, a welter of fictions that have shown every sign of getting the better of her—Rosalind is finally compelled by the sight of blood to betray her true womanly condition: Nature shows itself, and the play signals its approaching end. Or not. In fact, Shakespeare makes a point of having Oliver show Rosalind the bloodied handkerchief at the opening of the scene, at which point she betrays no response. In other words, it's not the blood as such but the narrative to which Oliver attaches it—the tale of Orlando's exploits—that prompts the swoon.

And, while the faint might be taken as a sign of Rosalind's falling out of the fictions, the response is prompted precisely by the fact of Orlando's tumbling further within them. After the encounter with the lioness and reacquainting himself with Oliver, Orlando leads his brother to his cave

> There stripped himself, and here upon his arm
> The lioness had torn some flesh away,
> Which all this while had bled. And now he fainted,
> And cried in fainting upon Rosalind.
> Brief, I recovered him, bound up his wound,
> And after some small space, being strong at heart,
> He sent me hither, stranger as I am,
> To tell this story, that you might excuse
> His broken promise, and to give this napkin,
> Dyed in his blood, unto the shepherd youth
> That he in sport doth call his Rosalind.
> *Rosalind faints*[38]

Rosalind apparently faints out of her role just as Orlando for the first time gives himself fully over to his, thinking even in his moment of extremis to convert the bloody signifier of his near-fatal encounter to the promissory token of his commitment to the terms of the game.

The threshold feeling of the exchange—the sense that one precipitates out as the other tumbles in—is implied by the imitative nature of the swoon itself. Perhaps Rosalind faints simply because Orlando faints. One might read that as a kind of unthinking somatic contagion cutting through the particulars of the narrative. Then again, Orlando doesn't faint from the wound; he faints after the event of the injury on seeing its evidence: It's the signs of the experience that precipitate the effect, just as it's the narrative of injury that prompts Rosalind's response. We're caught between soma and fiction, and what emerges from that suspended state would seem to be the proper name in all its irreducible uncertainty: "And he cried in fainting upon Rosalind."

The sense that affect is being aligned with moments of crossing over from one state or character to another is borne out by the tale Oliver recites, a conversion narrative, as it turns out. Whereas Rosalind and Orlando pass out, Oliver describes an experience of coming to. Though the audience is familiar with malign Oliver from the play's opening scene, he is a stranger to Rosalind and Celia. And although he now subordinates the entire account of his near fatal experience in the forest to the part it plays as a message in the game between Orlando and his beloved, that experience is for him one of personal transformation. He speaks of Orlando coming across a "wretched, ragged man, o'ergrown with hair, / [who] Lay sleeping on his back" under an old oak

(107–8). Who that man is only becomes apparent when Orlando recognizes him—"it was his brother, his elder brother"—but Oliver doesn't let his interlocutors know that he himself is that elder brother—that he's been telling his own story—until the moment he awakes from Orlando's battle with the lioness who had been lying in wait to attack the sleeper. "Twice did [Orlando] turn his back, and purposed" to abandon his brother,

> But kindness, nobler ever than revenge,
> And nature, stronger than his just occasion,
> Made him give battle to the lioness,
> Who quickly fell before him; in which hurtling
> From miserable slumber I awaked.
> Celia: Are you his brother?
> Rosalind: Was it you he rescued?
> Celia: Was't you that did so oft contrive to kill him?
> Oliver: 'Twas I, but 'tis not I. I do not shame
> To tell you what I was, since my conversion
> So sweetly tastes, being the thing I am.
>
> (128–38)

It wouldn't be hard to see this moment of self-announcement as a narrative ploy, Oliver's easy signaling of his utter break from an unconscionable past. But to read the rhetorical positioning in such instrumental terms is to fail to recognize its truth to the conditions of narration involved. In fact, Oliver wasn't present for the scene he describes—he was sleeping; the suggestion of solipsistic enclosure is conveyed not just by his somnolent condition but by the "green and gilded snake" that had "wreathed itself" about Oliver's neck as he sleeps, "her head, nimble in threats, approach[ing] / The opening of his mouth" (109–11). The very possibility of self-transformation depends, then, on giving oneself over to the narrative of the other, of telling oneself from elsewhere. The leap of faith entailed in that is emphasized by the fact that Orlando's account of gilded snakes and lionesses in the English forest would seem to be thoroughly fanciful—the intrusion of fabular romance into what had become a quasi-naturalistic pastoral.[39] Indeed, it could well have been that it was Orlando that violently fell upon him, as he had in the play's opening scene, a possibility that makes the ultimate cause of the all-important wound itself a matter of faith.

The scene plays, then, on an ethical imperative that preoccupies the drama from start to finish: The importance of not presuming the tale that absorbs you is the only tale, or even that it is ever your own. "This wide and universal theatre / Presents more woeful pageants than the scene / Wherein we play in,"

Duke Senior declares, a pronouncement Jaques misconstrues as authorizing him to describe in his famous set-piece that "universal theater" as if from an omniscient position beyond it (2.7.137–39).[40] Such an ethic is given point for the audience at the play's close. If we are taken aback by the sudden appearance of the god Hymen at the conclusion of a more or less naturalistic play—a deus ex machina where none is needed—we shouldn't have been: Rosalind had let it be known that she had been raised by her uncle, a "great magician / Obscured in the circle of this forest" (5.4.33–34). As we noted in the last chapter, we dismiss such an account—take it as a passing strategic deception on her part—only because we presume the story that has absorbed us from the start is the authentic narrative, foreclosing or subordinating whatever might suggest contradictory accounts and origins. Given the suggested allusion to the theater, there might have been good reason for us to have heard in the reference to the magical "circle" of the forest the more compassing site of origination for the diegetical tale that's held us, a source that nevertheless derives its originative authority from what fiction's forested reaches suggest of the illimitability of origins in the vast web of the world.[41]

But to read Oliver's conversion narrative simply in terms of the redeeming movement from one given narrative to another fails to account for the ambiguity of the moment of coming to he describes. "In which hurtling / From miserable slumber I awaked," Oliver says. The "hurtling" is the tumult of the battle with the lioness. But the word also harbors its more familiar verb form, tempting one to hear in it a description of the process of "hurtling" from sleep to consciousness. In that sense, the switch-word intimates another more subterranean possibility, that what the scene describes isn't just Orlando's account but Oliver's own dream narrative, a prospect suggested not just by the fabulous nature of what he describes but by the gliding, metamorphosing vantage point of the narrative itself, where one scene seems to unfold into the next. If Orlando's narrative contains Oliver, Oliver's at the same time contains Orlando.

Such an impossibly doubled perspective—what leaves us uncertain whether we've been inside or outside a space of consciousness—makes sense when we read the moment of passage between as at the same time a moment of autopoietic origination: not a particular instance of conversion from one self to another but an account of how the awaked "I" appears at all. Simply to be inscribed in one's own narrative would be to be as lost as the solipsistically absorbed and insensate sleeper lying beneath the oak. Consciousness as such appears only with the breaking of that circle by virtue of the caesura that lets one comprehend it as a narrative at all, a fact that makes any account of what preceded such a break a retroactive fiction—if we feel that we've slipped into untethered romance at this point, that may in fact be because the scene is gravitating toward, rather than away from, the grounds of consciousness. In other words, the moment of affective transformation becomes what Nancy describes it to be: an ellipsis or point of passage, a turning beside, that is at the same time the origin of the world from which it slips away.

To view the scene as a scene of autopoietic origination transforms one's understanding of what it describes. Rather than representing the solipsism of an already given subject, the wreathing "green and gilded snake" "who . . . nimble in threats, approached / The opening of [the] mouth" of the sleeping figure evokes the autoerotic turning around through which the subject comes to and fails to come to itself, the mingled beauty and danger of the gilded snake suggesting the double-edged character of that erotic moment of self-divisive self-creation.[42] That the head of the serpent approaches the opening of the mouth suggests both the dire prospect of such self-completion—the self-swallowing abyss of touching touching—and the fact that such a recursive instance is bound up with the movement into language, including the language of the narrative itself, suggesting a constitutive crossing of limits between soma and word, nature and artifice, picked up in the descriptive features of the "green and gilded" snake. The snake in that sense figures the self-swallowing structure of the narrative as a whole, the way in which it divides itself in order to incorporate itself, each version of telling at once gathered into and split off from the other.[43]

In its more bitter inflection, which is to say, insofar as the heteronomy of the reflexive conceit is experienced in relation to the supposition of an integral and thus threatened self, the gilded snake chimes with the opaque wellspring of violence in the drama: envy, the poisonous condition prompting Oliver's, as well perhaps as Orlando's, Cain-like animus. "Gentle" Orlando is "so much in the heart of the world, and especially of my people . . . that I am altogether misprized," Oliver had told us (1.1.157–59).

To understand why envy should appear at the animating core of a romantic comedy and why it remains an enigmatic cause—"for my soul—yet I know not why—hates nothing more than he," Oliver declares—we need to recognize the affect's relation to the elusive grounds of sociality and thus to comedy in its intrinsic enactment of those grounds (1.1.154–55). When Augustine traces the origins of human sinfulness, he cites the infant's refusal of any check on its insatiable appetitive demands but lingers more pointedly over the case of the little baby whom he himself observed, one who "before it could speak, gazed with a pale and bitter look at another child that sucked away its milk from it" (*nondum loquebatur, et intuebatur pallidus amaro aspectu conlactaneum suum*).[44] Envy embodies primordial sin as a version of sociality whose divisive mediatory taint afflicts the very source, arising before speech, at the mother's breast, striking the infant from the outset with the pale mark of mortality.

For Freud as well envy is at the wellspring of social formation. "The social sense," he writes, in terms that bring to mind the trajectory of Shakespeare's play, "is based on reversing an initially hostile emotion to become a positively stressed attachment that has the character of an identification," the rivalrous competition between brothers resolving itself in a mutual identification in relation to a common object of love, the idealized figure of the father—a dead father, insofar as he was himself the victim of the sons' prior envious violence.[45] Envy in Freud's tale of origination becomes a reciprocal and therefore oddly sourceless cause: Though Freud claims the sons' envy is directed against the father's apparent self-sufficiency and power, he also suggests that the father's suppression of the sons was itself driven by jealousy.[46] That ambiguity around causation is consistent with the larger contradiction entailed by Freud's account as a story of origins. The passage to collective identification originates in the sons' individual envy of the father. At the same time, "individual psychology," Freud insists, is first constituted as an outcome of the agonistic drama, that drama amounting to the condition for the possibility of identification at all.[47]

Envy crystallizes that sourcelessness as the irreducible condition of social formation. "Original envy," as Freud terms it, figures identification as identification not with the other but with the place of the other, a matter of "wanting to put oneself in the same position."[48] Rather than being a matter of subjective desire or objective identification, envy bears on the more primordial positing of the relational structure that institutes such oppositions. Constitutively mediated, envy conveys desire itself as formed via its displacement, as the desire of the desire of the other, that is, at once a desire for the other's desire and a desire from the place of the other.[49]

The division implied by that originative vicariousness—a splitting Jacques Lacan associates with the emergence of the subject as subject of the unconscious—is realized in envy's distinctively self-consuming character, as that which is formed at once through and against the other and thus through and against itself.[50] Envy's relation to the forest scene in all its reflexive turnings is suggested by Giotto's allegorical depiction of the sin in the Arena Chapel, where a snake emerges from the wizened figure's mouth looping back to threaten its own envious eye, evoking the self-poisoning vice's relation to the abyssal character of allegorization itself (see Figure 1). "The envious man is consumed inside and out," writes Johannes Viterbiensis—consumed inside and out because, like the self-consuming forest narrative, it threatens the distinction between inside and out, self and other.[51]

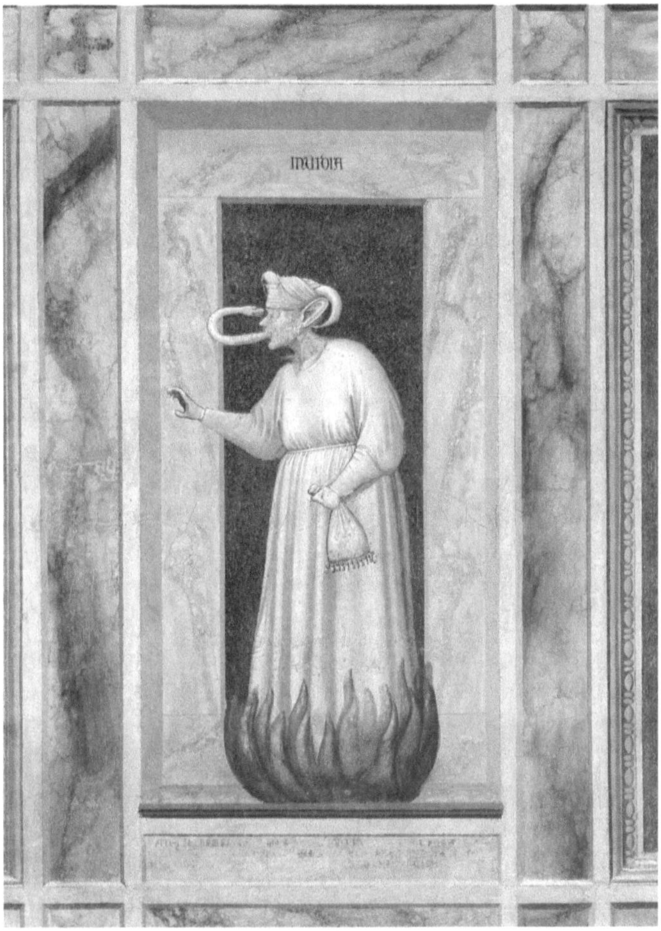

Figure 1: Giotto, *Allegory of Envy* (Invidia). Fresco. 120 x 55 cm. Scrovegni (Arena) Chapel, Padua, Italy. (Mondadori Portfolio/Hulton Fine Art via Getty Images)

Rather than relating internal or external events beyond itself, Oliver's narrative should ultimately be seen to allegorize—at once describe and perform—its own elliptical self-creation. All that follows in the narrative can be read as the unfolding of that initial autoaffective spiral as it moves by "indented glides" recursively upon and beyond itself, opening worlds in the process. "But suddenly" the snake,

> Seeing Orlando, . . . unlinked itself,
> And with indented glides did slip away
> Into a bush, under which bush's shade
> A lioness, with udders all drawn dry,

> Lay couching, head on ground, with catlike watch
> When that the sleeping man should stir.
>
> (4.3.111–17)

The unlinking, linking narrative unfolds as if by an associative logic of its own, one scene gliding into the next as though without need for a governing consciousness. And it does so because it tells in its process the story of the rivened emergence of such speculative consciousness in the movement from orality and blind touch to the gaze to the moment of coming to. That the animal gaze conjured in the process is at the same a reflexive gaze is intimated by the fact that the maternal lioness can't move until its object does, suggesting an uncannily divided origination, consciousness as a matter of seeing itself unseeing; the fact of environmental entanglement registers in that regard as one's subjection to a gaze as intimate as it is alien.

"This seen," the narrative then modulates into the moment of Orlando's recognition—"it was his brother, his elder brother"—but it's a fraternal cognizance that does nothing to arrest the indented turnings: "Twice did he turn his back" (128). The "hurtling" that follows—whether salvific or homicidal, a turning toward or from companionship—leads to the moment of coming to, the appearance of a consciousness realized only by virtue of ceaselessly dividing and returning on itself: consciousness as such as the "turning" of "conversion."[52]

By the logic of its breached creation, such an autopoietic, internally self-constituting narrative can equally be seen to originate from beyond itself, as a function of its interruption—the formative internal/external "irritation" of autopoietic institution. It is at each turn Orlando's own intruding presence that sets the scene in motion—his presence that prompts the serpent to "slip away," his intervention that rouses the lioness. The severing completion that marks the snake's autoaffective touch equally informs the structure by which Orlando's interruptions produce the scene within which he is at the same time gathered. The scene first comes to view as an effect of Orlando's own turning aside. Oliver begins his narrative by describing Orlando:

> pacing through the forest,
> Chewing the food of sweet and bitter fancy,
> Lo what befell. He threw his eye aside—
> And mark what object did present itself:
> Under an old oak . . .
>
> (4.3.101–5)

The object "presents itself" even as the language of the passage—"lo," "mark"—emphasizes the determining force of the gaze: Twice over we are enjoined to look—to look at his diverted looking—eliding the distinction between onlooker and scene and leaving it unclear at what point we've entered the scene. For Orlando as well the beautiful and threatening scene he encounters through his turning eye is at once a fantastic new world and a reiteration of his own "sweet and bitter" condition, the "fancy"—or fantasy—he, like his self-consumed brother, "chews." Orlando's aversive turning, his "thr[o] w[ing] his eye aside," like his "twice . . . turn[ing] his back," already predicts the self-exorcism through which Oliver becomes himself. An autoaffective turning that exceeds as it completes itself, a breaking-in that constitutes the world into which it intrudes: If each brother is finally able to recognize his fraternal other in the scene, it is insofar as both are already and from the outset beside themselves, both inhabitants along with their creatural companions of a space that is itself incapable of totalization and without apprehensible source—their own world and one always "more than the scene wherein [they] play."[53]

Beginning neither within nor beyond, the scene originates—produces itself as a scene—instead as the fantasized trace of its own aversive turning away. The torsional logic of such world creation is implicit in the figured action that amounts to the allegory's enigmatic kernel, the violence that at once fissures and enables the narrative's ouroborian self-incorporative self-creation. In its noun form, a "hurtling" is a conflict or violent clash. To be "in" a hurtling is to be within a space of conflict. In its verb form, it is a violently oppositional movement—a dashing or hurling against. Condensing those meanings, the term in the context of the scene suggests not just a passage between spheres but a divisive movement that constitutes the very space in which it unfolds.[54] The term in that sense condenses the ellipsis of world creation as the scene enacts it, implying a "cause" that remains unknowable insofar as it determines the space and condition of its apprehension. Understood thus, "hurtling" is another name for affect—the term derives from "hurt"—suggesting that the injury on which the scene hinges—the reason for the message Oliver conveys as well as the source of its power to afflict—is neither a literal one nor an inward psychic injury but the wound that, opening such dimensions in the first instance, materially and inapprehensibly inscribes.[55]

How affect wounds, and thus how it bears on theater's hold, is a function of its temporality. The recursiveness of autoaffective touch means that affect will inevitably be precipitous—the language of falling and befalling runs throughout the passage—insofar as it begins at the point of its returning, arrives ahead of itself. That anticipatory property, its coming from the future, as it were, is of a piece with the transferential nature of the responses it provokes: Orlando's turning foretells Oliver's; Orlando's faint, Rosalind's.

To read such effects as mimetic or imitative is to lose the source of their force and, indeed, what makes them affective at all, a truth we can recognize by acknowledging the extent to which the scene of fainting implicates us, how much it has been from the outset the effect of our own intrusive desire to "mark" it. For the audience, the uncertainty about how to construe Rosalind's lapse—truth or, as she claims, "counterfeit"—has the structure of a trap, the familiar net of autopoiesis. To claim it as the "feminine" truth of the matter, to claim that we see the limit of the game playing and with it an anticipation of the play's own end, is of course to fall into the fiction perpetuated by the boy actor who plays Rosalind: We're caught within by virtue of claiming a place beyond. That ensnarement isn't merely thematically contingent or remediable. To "mark" the fiction at all—to make any fiction knowable *as* fiction—one must draw the line, break the circle, and thus determine one's place beyond it; that is the condition of our "coming to" as subjects at all, as Oliver, turning away, becomes what he is. And yet in that aversive turning away—our claim to a place beyond—we merely reiterate a scene constituted from start to finish by such a turning aside: It is thus that fiction seduces and inscribes. Fainting Rosalind in that sense amounts to the figure of our own knowing and insensate absorption. The tropic space of fiction and of affect has never been anything other than that expansive turning in and beyond itself.

Such captivation isn't mimetic; it's inscriptive. Turning aside we actively constitute the scene as what was already there, what already binds us.[56] The temporal syncope of such a self-divisive origination gives it the force of a wound. In its elemental form, affect amounts to a flight that constitutes what it at the same time flees, fainting the refusal of affect as affect's origin: thus affect's inescapable contagiousness.[57] Just as according to its autoaffective logic Rosalind's faint can be said to be caused by the fact of its being caused by Orlando's and vice versa, our response at such a theatrical moment is caused by its own vertiginous causelessness. With that, affect takes on its defining character as an interpretive response in excess of any cognizable object, even as, in thus insistently exceeding the frame of its own formation and of our own capacity to confidently situate it there, it constitutes the truth of reference as such—naming in its irreducibly proper and questionable form: "And [he] cried in fainting upon Rosalind." *Pace* Luhmann, such an indicative force can neither be securely described as auto- or heteroreferential, internal or external.

Understood in these terms—as a causeless cause—the hurtling force evoked in the scene has a cognate in Heidegger's thrownness, the propulsive condition and event of being or *dasein* insofar as it is always already a responding and thus never separable from our positing it as a question. In that regard, the scene figures world not as a domain of objects but, to follow Heidegger's account, a more primordial condition of responding and care.[58] "What occurs" by virtue of affective being's constitutive finitude and noncoincidence, "what bends, leans, twists, addresses, denies—from the newborn to the corpse—is neither primarily 'someone close,' nor an 'other,' nor a 'stranger.' It is an origin"—"the discreet passage of other origins of the world."[59]

To put this another way, affect is what bears out being's irreducibly historical character, for that opening of always other origins and other worlds amounts to the fact of historicity as such. The division and displacement that hollows out the "now" of affect, that determines it as a missed encounter or one still to come, is the realization of the temporal contingency of being, its status not as essence but as event. And that event—or eventless event—is the fact of our own historicity. Constituted in its interpretive marking, the play comes to or realizes itself through our present reception, even as that reception proves itself already inscribed and foregone, the historicality of the play thus asserting itself not as a story of causes but intimately, within and as the torsional wound of affect itself.

Considered in these temporal terms, the scene's hurtling passage—an affective flight from and a coming to—resonates with another allegory, this time a familiar modern one posed by Heidegger's own fratricidal rival. Walter Benjamin's angel of history—the famous figure from his "Theses on the Philosophy of History"—also conveys a story of hurtling forward and turning back. "Where we see a chain of events, he sees one single catastrophe, which keeps piling wreckage upon wreckage, and hurls it before his feet."[60] Even as the angel would turn back to "stay timeless time" and undo its damage, the storm of that catastrophe—another name for "progress"—"irresistibly propels him into the future toward which his back is turned."[61] Again, the historicity the scene describes can't be separated from its witnessing. The angel, who would stand outside time to contemplate it in its timeless wholeness is held fast within the storm's winds precisely by the fixed horror such a vision provokes; "look[ing] as though it is about to turn away," the figure nevertheless "fixedly contemplates" what he beholds, "his eyes are staring, his mouth is open, his wings are spread."[62] And what he beholds—"a storm blowing from Paradise"—is us. Klee's *Angelus Novus*—the image from which Benjamin derives his allegory—in fact shows a frontal portrait, its open-mouthed expression reflecting the viewer's own surprise or horror or insensate vacuity—a figure, then, of our own deadening flight and historical coming-to. History coincides with the autoaffective logic of affect, a propulsive and exorbitant condition insofar as it is the event on which it reflects. In that regard, hurtling is another name for the trauma of worldly immanence.

6

The Spider and the Spinning Wheel

Techne, Animality, and the Autopoietics of Touch in Velázquez

No visual artist, one could reasonably propose, has played as central a role in the development of modern theory as Diego Velázquez, beginning with Michel Foucault's seminal analysis of *Las Meninas* at the opening of *Les mots et les choses* and Jacques Lacan's account of the same work in his seminar on the object in psychoanalysis. Such prominence in the theorization of subjectivity, representation, and history is hardly surprising given the self-consciousness with which the painter's works reflect on their own representational status, the consistency and clarity with which they frame their own endeavor. Velázquez's art is, from the outset, a theory about art. At the same time, surveying that critical literature from the era of high theory, with its largely representational orientation—its concern, for instance, with matters of perspective—one might be struck now by how unengaged it was with the quality for which the painter was best known: his painterly touch, and the uncanny sense of pictorial immediacy it was able to conjure. "The most painterly painter who ever lived," writes Theophile Thore in 1857; the "painter of painters," according to Manet.[1]

That contradiction—an art of self-consciously flagged artifice and of sensible immediacy—should be familiar to us: It is the hallmark of autopoietic formation. Indeed, Velázquez lets us complete the itinerary of autopoiesis as missed encounter from the world-dissolving pathos of early modern tragedy to the most refined precincts of reception aesthetics. Kenneth Clark famously describes repeatedly stepping from a distanced, clarified view of a Velázquez painting to a closer view in which the work dissolves into a profusion of apparently random brushstrokes without ever being able to catch the moment of transition, which is to say, without being able to capture the moment the image becomes an image—a moment as elusive, he says, as the transition from sleeping to waking.[2] What Clark describes is simply the performative mystery of autopoietic origination and the elliptical knowing unknowingness it underwrites, the enigma of Caliban's waking dreaming.

Limning that aesthetic condition with particularly rigor, Velázquez helps us locate the self-consciously "literary" turn we've traced especially in Shakespeare's later works within the broader Kantian or proto-Kantian transformation that occurs between the Renaissance and the nineteenth century from an art of the object to an art of the perceptual act, from mimesis to phenomenality, from "the image as lie," Jean-Luc Nancy writes, "to the truth as image."[3] By exposing the conditions of that process of aestheticization, Velázquez's work poses the ever-challenging question of the relation between autopoiesis and history, including the kind of explanatory genealogy just adduced. That tension between structure and history ultimately manifests as a contradiction within autopoiesis itself, an aporia between phenomenality and reference that crystallizes in the painter's touch; by a logic we have already intimated, it is within the subtilized domain of the self-consciously artful work that the problem of reference, and with it the problem of world, is most acutely posed. Evoking a force of reference necessarily outside the process of phenomenalization it enables, Velázquez's art adumbrates the possibility of thinking the world other than as mimetically conceived, other than as a world that lies before us, and even beyond the consolidating autopoiesis of aesthetic consciousness. As we will see, that means reckoning with an inhuman dimension—the mechanical turning of the spinner's wheel, the opaque gaze of the animal—not as an element in the world but as the world's condition.

1

Given the painting's complex metacritical conception, it is hardly surprising that the bulk of the theoretically inclined writing on Velázquez has focused on the late masterpiece *Las Meninas*. I will turn to that work and the critical literature it has spawned in the latter part of the chapter. But I want to recast our understanding of that image via a detailed consideration of Velázquez's other late, large-scale piece in the Prado labeled variously *Las Hilanderas* (The spinners) or *La fábula de Aracne* (The fable of Arachne), a work no less ambitiously reflective of its own status as art but that figures those concerns via something like an allegory of touch—Giles Knox speaks of the painting as a "celebration of the hand"[4] (see Figure 2).

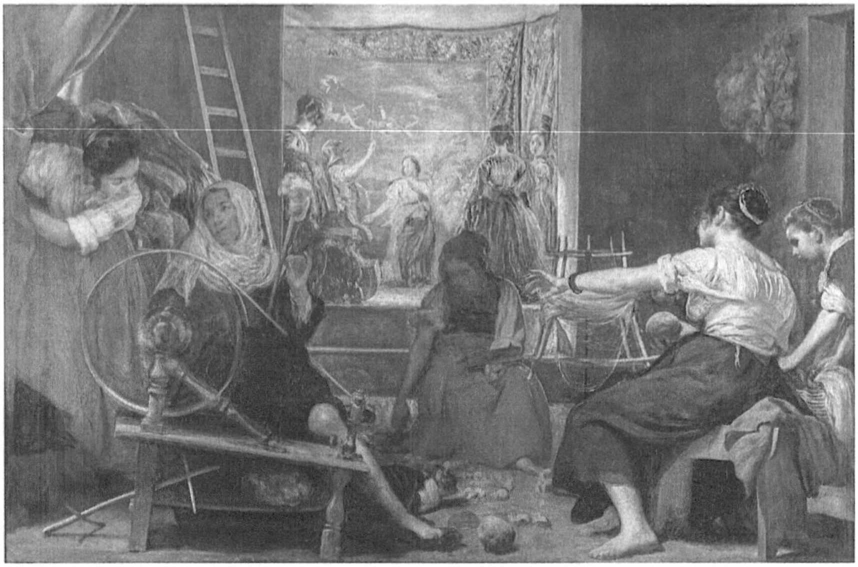

Figure 2: Velázquez, *Las Hilanderas*. Oil on canvas. 220 x 289 cm. 1655 (Museo Nacional del Prado. Inventory # P001173 ©Photographic Archive Museo Nacional del Prado)

The confusion about the painting's title arose because the subject of the inset scene—Minerva's moment of judgment against Arachne—wasn't recovered during the modern era until the 1940s. Understood during the painter's own time as an allegory of artistic agonism and progress,[5] Ovid's tale of the goddess's envious judgment against the upstart girl reveals its artistically self-conscious status here in Velázquez's rendering of Arachne's finished tapestry in the background of the inset scene. The first of the girl's renderings of the gods' scandalous misdeeds, the representation of the rape of Europa is in fact a reproduction of Titian's painting of the scene, then in Phillip IV's Escorial collection (see Figure 3). Velázquez stakes his claim in relation to the Renaissance artist most esteemed by Spanish painters of the era, a painter himself known for painterly touch—Titian's painting had inspired a series of copies, including by Rubens, the most famous of Velázquez's contemporaries, during his stay at the Spanish court.[6]

The paragonal nature of Velázquez's work is apparent in the inverted hierarchical oppositions it represents: between the individualized accomplishments of high art reduced to the background and collective mechanical labor in the fore; between *istoria*, the most esteemed mode according to the era's norms, represented in the mythic tale, and genre painting elevated by scale and by the spinners' classicizing postures; between the conceiving mind of the male artists and the unthinking female hand figured here in its kinetic

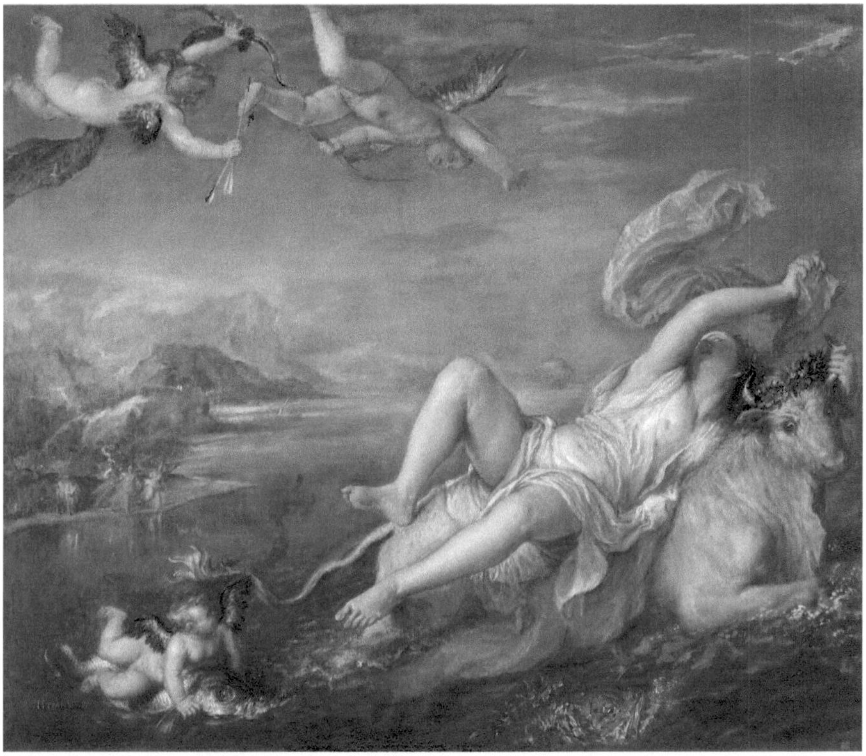

Figure 3: Titian, *Rape of Europa* (By permission of the Isabella Stewart Gardner Museum, Boston)

movements as an arachnidal digital multiplication.[7] But the painting's status as artistic display piece—its claim to outstrip Titian—is most directly conveyed in the spinning wheel, a detail that explicitly responds to a challenge set down since antiquity with Pliny's claim that the painters Antiphilus and Aristides were able to depict motion in the static medium of painting, the latter's representation of a four-horse carriage "so lifelike … that one would swear that the wheels turned and rolled along."[8] By some account the first modern painter to reproduce such an effect of static motion, Velázquez here creates it via a stroboscopic evocation of the limits of perceptual grasp—a form of conjuring effacement.

The painting's character as a reflection on artistic inheritance is in keeping with the myth itself. For Arachne's transgression—what precipitates the confrontation with the gods—isn't her claim to superiority; it's her insistence on her singularity, her claim to an art without forebear—and especially that she had learned nothing from her divine predecessor.[9] The foundational nature of that act of artistic repudiation, whether Arachne's or Minerva's, is picked up in

Velázquez's unusual choice for a moment to depict: not the scene of punitive transformation, the moment conventionally favored including by Rubens, but the prior moment of judgment conveyed here in Minerva's raised hand, a gesture that reads as acclamation as much as castigation. It was the goddesses herself who judged the competition, and her envious violence against girl and image—she tears Arachne's tapestry—amounted to an acknowledgment. That sense of a suspended gesture is apt, for the god's intervention was, in the end, as protective as it was punitive, a reversal of her initial fatal judgment that leaves the girl herself hanging: Taking pity on the girl at the moment she is about to hang herself in despair in the face of the goddess's wrath, Minerva transforms her into a spider, leaving her, perhaps more cruelly, like art, forever suspended.

The logic of affirming negation is echoed in the divisively autoformative economy of the painting, the way it defines itself at once as and against high art, incorporating the space beyond art's frame. Velázquez, it has been observed, had an ambivalent relation to the labor figured in that quotidian space beyond. The artist was acutely conscious of Spain's belatedness in according painting the status of a liberal art, and he was preoccupied with attaining a knighted position in the royal court, a role he assumed at the cost of artistic output.[10] At the same time, it was on art as the work of the hand that he would stake his claim. Here, the apparent contradiction underwrites the supercessionary logic by which the painting translates the opposition itself into the grounds of a higher art. Thus the double meaning of weaving in this context, its functioning at once as one term within the opposition of high and low and as a figure for art's prior ground or material substrate—consider in this light Velázquez's technique of incorporating exposed canvas often to dazzlingly ambiguous effects of distance and proximity.

The painting's inclusively exclusive economy extends from its broadest historical instantiation to its most finely tuned sensory effects. Of a piece with the range of tropes figuring the West's origins by way of a moment of female abduction and sacrifice—Helen, Dido, Cleopatra—and consistent with the Arachne myth's early modern allegorization as the triumph of modernity over antiquity,[11] Tyrean Europa's rape signifies Europe's forced appropriation of an African or Middle Eastern prehistory but equally the internal division by which Europe constitutes itself from the outset in its autonomous status as disinherited heir.[12] That contradictory internal/external determination is reflected in the Ovidian myth's notable gentling of the boundary between seduction and abduction, the docile bull often read as a figure for aesthetic absorption[13] and the abduction itself graded to the point of imperceptibility as the creature "rises ever so gently and slowly edges / From the dry sand toward the water, further and further," with Europa clinging to its horn for safety's sake.[14]

Such gradualism defines Arachne's weaving as well. "The threads that touch seem the same, but the extremes are distant," Ovid writes.[15] Here too, even as a purely aesthetic matter, nuance is bound up with force, a force that amounts to the condition of phenomenalization as such. Velázquez's wheel figures the movement by which perception outstrips itself as a specifically reflexive mechanism, condensing at the level of sense perception the larger mechanism through which the image as a whole recursively at once constitutes and negates its own artistic grounds, just as every figure as a figure forms itself at once through and against a posited ground.[16] Clark's experiment—his performed inability to locate the image's emergence in space and time—simply reiterates the enigmatic source of the painting's phenomenal formation, its realization of sense as the seductive abduction of sense.[17] The feeling of movement is in that regard something more than a painterly device: Such temporal precipitancy is the condition of the painting's emergence as aesthetic formation, a matter of the image exceeding its perceptual frame as the condition of its originative enframing.[18]

That aesthetic logic is consistent with an apparent contradiction that has marked the reception of Velázquez's art. On the one hand, the paintings are noted for their reflexive self-consciousness as a medium. Knox has written at length of the extent to which the works reflect on the mechanism of their own production, from Vulcan's tools of the trade, to the weaver's craft, to the foregrounding of the act of painting itself in *Las Meninas*.[19] That highlighting of means extends, of course, to Velázquez's outright display of brushwork as brushwork. At the same time, the paintings are celebrated for their experiential immediacy, the effect they produce of unmediated presence.[20] The impression of a hand and no hand recurs: "It seems as if the hand doesn't take part in the execution but it is painted only with the will. Of this kind, it is a singular work," Anton Mengs writes in 1797; "his painting is almost done without painting," Svetlana Alpers more recently remarks.[21]

To make sense of that phenomenon of unmediated mediation we need to recognize how the turn from mimesis to phenomenality, what Maurice Blanchot describes as the historical appearance of "the power of the work to be and no longer to represent," also means a translation of the image from the status of object to a form of seduction or contagion.[22] When Minerva raises her hand in condemnation, she bears out precisely what she seeks to repudiate in Arachne's tapestries, their depiction of the arbitrary violence of the gods. She inscribes herself, just as Velázquez inscribes Titian and Ovid embeds tales within tales in the endless and endlessly seductive weave of his narrative; indeed, the scene of judgment seems to merge with the luminous tapestry behind, as distinct from the foreground. That prospect of framing oneself by virtue of the force of one's framing judgment—the hallmark of the work's status as reiterative autopoietic form—is played out in the instability of the identificatory oppositions the painting mobilizes. Do we align Titian with Arachne—it is his image after all—and thus Velázquez with the enviously self-exposing Minerva? Or do we identify Velázquez with upstart Arachne, who bears out her imitative dependency precisely in the fierceness of her refusal of that inheritance? In either case, it's the act of judgment as such, the gesture by which one presumes a place beyond the frame, that inscribes. Painting is absorptive, in that sense, precisely to the extent that one insists on its status as mere art.

That logic of the image as lure plays out in the visual dynamics of the painting. Signaled as a space of voyeuristic penetration by the figure in the foreground drawing back the curtain on the scene, the painting leads the eye along the frontal plane from right to left, following the line indicated by the weaver's extended gesture and then slanting back into the recesses of the image, guided by the focalizing diagonal of beams of light. In our claim to comprehensive seeing, as well as to interpretive knowingness vis-à-vis the allegory, we are situated in a position parallel to Minerva, who, with her back to us, pronounces judgment on the image within. And yet, just to the extent that we claim our power as unseen seers thus, to the extent that we subjectivize ourself in relation to the inanimacy of the image as object, we will have been caught out by a gaze that returns to us from just beyond the boundary of our penetrative sightline, the gaze of a female figure in contemporary dress whose look cuts across or short-circuits the painting's framed recession. As for Minerva, the viewer's force redounds, an effect whose power to inscribe is directly connected here too to its temporality: The painting has foreseen us.

Whether at the level of historical framing, the structure of the gaze, or the mechanism of phenomenalization, such inscriptive effects play out according to an autopoietic economy that, though illimitable, is fully cognizable, an economy of negation that systematically reverses absence and presence, subject and object, resistance and capture, agency and subjection. In that regard, the painting can be taken to exemplify the historical emergence of aesthetic consciousness, a distinctly modern condition whose claim to an illimitable capacity to form itself in relation to its own limits—to claim a groundless ground—sorts with the painting's self-exceeding structure, the way in which it constitutes itself by at once including and breaching its own framing limits.[23] The redounds entailed in such an economy bear out the intimate complicity between aestheticization and law, divine or otherwise.

But details of the painting also suggest the way in which aestheticization shades into less systematic effects, the way in which the process that would translate autopoietic origins into knowable form ultimately troubles the logic of autopoiesis itself. To speak of the detail only as an index of artistic virtuosity may overlook what is most striking about the spinner's wheel: the sense that it operates of its own accord and thus outside the entire opposition of hand and mind, labor and cognition. The implication such a prospect holds for economy generally is evident in Aristotle's account of the role of techne:

> If every tool could perform its own work when ordered, or by seeing what to do in advance, like the statues of Daedalus in the story, or the tripods of Hephaestus which the poet says "enter self-moved the company divine";—if thus shuttles wove and quills played harps of themselves, master-craftsmen would have no need of assistants and masters no need of slaves.[24]

Especially insofar as it is bound up with the most fundamental conditions of sense experience, the integrative or sublating vocation of aestheticization is directly related to the means through which a potentially sheerly dispersive activity is gathered into a coherent productive totality, the process through which labor becomes end-oriented labor and economy is constituted as a system of coherent returns. Like Aristotle's statue of Daedalus, and like the statue at Mitys, Velázquez's spinning wheel suggests that even in fulfilling that recursively consolidating function, art may exceed the dictates of the productive hand.

That suggestion of a dimension beyond the compass of human agency is not confined to art's relation to material techne. We noted the gaze that returns from just beyond the image's focalized recesses, playing off the painting's frame logic. But to stay within that organizing sightline and follow it to its overexposed limit means encountering another gaze: the eye of the white bull in Titian's painting. In narrative terms, the bull is the aggressor, yet it is hard not to read that gaze especially in Titian's rendering as one of mute pathos, as much the look of sacrifice as of predation. Is it the eye of the god or of the animal? What becomes of the animal when the animal is possessed, made pure vehicle whether by gods or by art? Sacrifice, civilizations' founding violence, reaches a recessional limit in the abduction and loss implicit in the movement of metaphorization itself, the sense of a remainder lingering within and beyond prosopopoeiac figuration.[25]

Situated at the image's vanishing point, a contradictory point here of greatest recessional depth and of greatest overexposure, the divided eye of the creature corresponds with our own gaze at the point of its convergence with image, suggesting not the reflective or returning gaze of the painting—not a symmetrical reversal of subject and object—but a more thoroughgoing conflation of seeing and painting: sight as blind spectacle—recall the uncanny presence of the watching creature in the Forest of Arden. Absent the logic of the frame, and thus beyond the sacrificial economy of negation that underwrites aesthetic consciousness, the animal gaze implies undialectizable distance as proximity—the exposure without depth of an intimate exteriority. If the image catches us at that level, it is not according to a logic of recuperative negation but by virtue of one's originative and inassimilable foreignness to oneself.

Associated with techne and animality, art's relation to the limits of human agency is perhaps most profoundly figured in the action of the creative hand itself. As against the claims to originality implied by high art, claims that merely conduce to rivalrous reversals of power and dependency, the older weaver's upraised hand suggests an explicitly mediatory conception of the creative act—a matter of guiding the thread that extends beyond it. At the same time the gesture echoes the movement of the spinning wheel in its own rotated form. Compressing those dimensions of the image to a perceptual limit, the weaver's hand suggests a generative source more fundamental than the opposition between mediatory division and self-sufficient origination. The sense of divided autonomy conveyed in the painting as a representation of a unitary community of women riven by an act of judgment condenses to an evanescent point in an encircling gesture dissected by the sheerest thread, nothing more than the luminous trace of paint itself; the reflexive, metapainterly dimension of the gesture is picked up in the way the suspended hands of Minerva and Arachne, one closed, one open, are at once bisected and joined by a vertical line of white paint indicating a fold at the midpoint of the tapestry and the painting. The painting's larger self-inclusive logic crystallizes as a virtual self-touching in which the distinction between touching and touched approaches a dissolve, implying an autonomous/heteronomous origin as enigmatic as a spider's self-spinning web. Recalling Montaigne's female spinners in their troubling of the opposition between techne and human agency, production and affective pleasure, the gesture intimates art's relation to a radical undoing of economy and productive aim and thus the possibility of thinking art and community otherwise.

The painting's relation to the animal dimension may also be conveyed or enacted in just such a dispersive unbinding of the work's own economy as a thematic or end-directed form. Notably unrepresented in the painting, Arachne's defining transformation is nevertheless present everywhere, one could argue, the spider's multiplicitous touching realized in the centrifugal profusion of details, as well as in a fugitive brushwork indistinguishable from the scattered movement of the eye. Such nonfiguration may be just to the point. We have described Velázquez's touch in terms of an aesthetic of the missed encounter—Kenneth Clark's inability to capture the image's moment of appearance, the concentrated indistinction of the spinning wheel, the impossibility of discerning differences traced across the length of Arachne's tapestries. Such moments give form to temporal and cognitive indeterminacy precisely within the terms of a consolidating temporal experience, an occurrence *within* time—that temporalization of loss is precisely what makes aesthetics a matter of aesthetic consciousness. The same presumption of time as an already given dimension lets Luhmann conceive autopoiesis as a systematic play of openness and closedness, inclusion and exclusion—as a coherent *system*.[26]

But such a temporalizing conception stakes itself against a more radically dispersive and multiform prospect of otherness, not the returns of the eye or wheel but a weave or web in which the distinctions between eye and painting, painting and material substrate dissolve. Velázquez's free brushwork "offers no univocal message ... but brings the viewer up against the ineluctably differential and an-archic character of perceptual and interpretive coherence," Veronique Foti writes.[27] With that disarticulation of object and perception, touch becomes the diffraction of touch, a matter of the simultaneously closed and open hand.[28] The version of community implied by such apprehension, an untotalized community of being as being-with conjoined in the very movement of its dispersal, can be intuited if the spinners' gestures, detached from their enlistment in the painting's thematic aims, are seen to articulate their own signifying chain, a kind of secret rapport implied in the way the women's variously opened and closed hands answer each other across the painting's intervals.

Traced thus beyond the temporalization of consciousness, Velázquez's touch implies a relation between aesthetics and history more fundamental than the sequencing of art historical epochs. "The deepest ambition of aesthetics," Marc Redfield writes, is "to gather the referential insistence of 'history' into the totality of consciousness or form."[29] *Las Hilanderas* shows the means through which that aesthetic appropriation occurs, for the divided reflex by which the image forms itself through and against its own ground, whether that included exterior refers to material conditions or the work's artistic forebear, is the mechanism through which context as such is instituted; narrative history is in that sense an autopoietic construction. The painting also suggests the force entailed in that assimilative process in its association of art's self-formative act with civilization-forming violence, an abduction in which infinite nuance spurs infinitely expansive returns of punitive force and where law feeds off

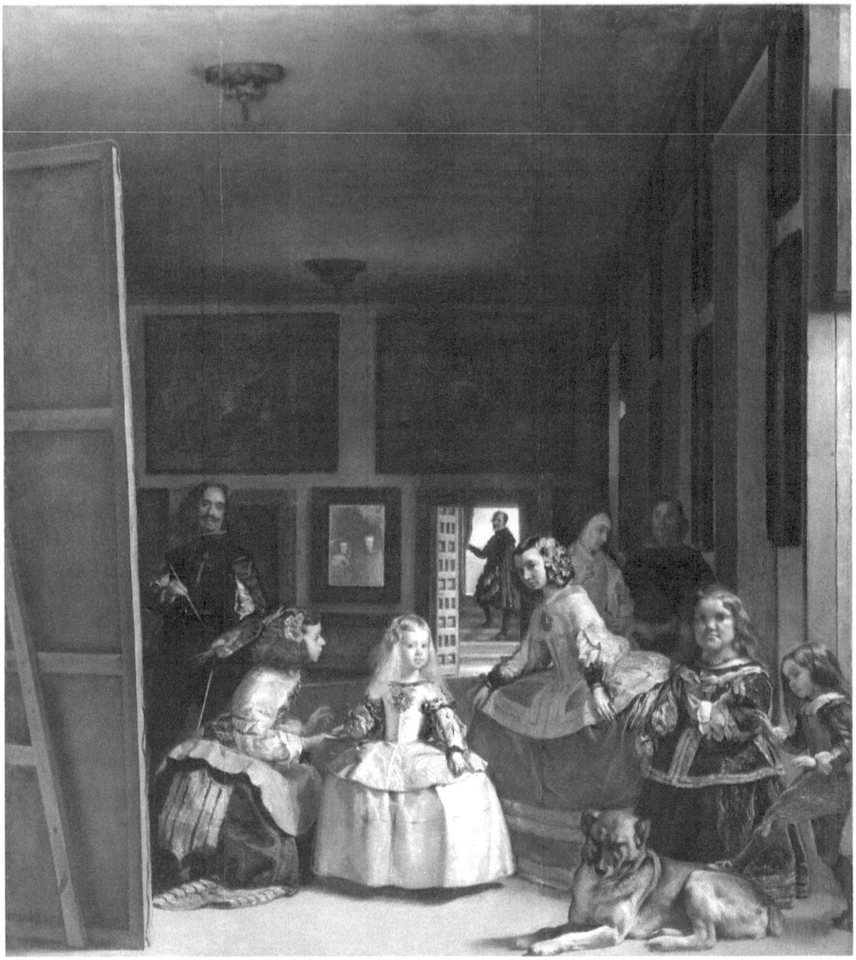

Figure 4: Velázquez, *Las Meninas*. Oil on canvas. 320 279 cm. 1656. (Museo Nacional del Prado. Inventory # P01174 ©Photographic Archive Museo Nacional del Prado)

its own groundlessness. In this light, Velázquez's twin allegiances to demotic representation and to the imperial court may not be such a contradiction.

At the same time, precisely as a consequence of its self-referential intensity, the painting adumbrates another relation to history, one associated with the exorbitance of reference as such. As allegory of artistic origination, the painting with all its recursive self-involutions—the turning wheel, the encircling hand, the redounds of the forms of judgment it describes and provokes—can be seen to emerge from the painting's most irreducible point of self-reference, the point where the painting as diegetic narrative incorporates itself as prior material condition, figured here in the luminous line of paint threading through the spinner's hand. The necessity and the impossibility of that self-divisive self-reference suggests a contingency inassimilable to the logic of autopoietic contextualization or aesthetic consciousness and thus resistant to the reversive economy of judgment and law, even in the suspended form that judgment assumes as aesthetic judgment. In *Las Hilanderas*, that referential drive remains minimally contained within the narrative that encircles it—a story about art. Something of the more radical and outright character of the relation to history in *Las Meninas* can be adduced from the translation of the weaver's poised hand into the more overt self-reference implied by the painter's comparably suspended gesture. I want to turn now to that painting, an image that conjures itself out of a moment of interruption—an interruption of and by history—and in which the question of sovereignty hangs on the tip of a painter's brush (see Figure 4).

2

The pertinence of Velázquez's other late, large-scale painting for a consideration of autopoietics is evident enough in its complexly reflexive, self-framing, and notably self-negational form—a good third of the painting consists of the turned back of the painting on which the depicted painter works. That metacritical dimension has been at the center of the theoretical literature the work has spawned, work such as Foucault's analysis of the painting's complex perspectival structures. I'll suggest, however, that the painting's force as autopoietic formation, and with that, its relation to history, only becomes fully apparent at the limit of such representational considerations, the point where phenomenal sight and language converge as touch.[30]

Placed at the opening of the volume, Foucault's analysis of the painting has itself something of the ambiguous status of a frame; it epitomizes one episteme among others—the Classical order situated by his account between the Renaissance and modernity—even as it oversees the book as a whole, engaging problems of representation as such and already intimating the turn to the modern era of the human sciences. The averted canvas lies at the core of that Classical order according to Foucault, for it is the occlusion at the heart of the image that sets in motion the ramifying substitutions that make up the Classical order as a signifying system; if the Renaissance functioned according to a logic of resemblance that bound word and world, the era of representation, as Foucault termed it, structured itself as a purely autonomous system of signs—a mathesis—that seeks its grounds within itself.

In the image, as for the era, what ultimately secures that system according to Foucault is the figure of the sovereign. While according to the perspectival sightlines of the painting as Foucault sees them the place of the actual painter, of the model, and of the viewer all converge, only the sovereign—or the sovereign couple king and queen—is reflected in the mirror placed directly opposite our viewpoint in the dark recesses of the depicted room. For according to the representational logic of the era, the subject remains the one term that can't be incorporated into the signifying structure, that must remain occluded for the system to become an objectifiable system. And yet the very tenuousness of that anchoring figure of the sovereign in the mirror—the sense of its fragility and adventitiousness—anticipates the turn through which, with modernity, the subject in all its finitude and contingency will be inscribed in the place of the sovereign, a turn that amounts to the introduction of the fact of historicity into the epistemological field. In sum, the liminality of the placement of the image in Foucault's analysis corresponds, one could argue, to a significant equivocation between structure and history in Foucault's account: What would seem to be a purely structural condition—the fact that no system of representation can become a system except by virtue of the exclusion that frames it—instead becomes an epochal contingency hinging on a subject who might or might not be incorporated.

While critical accounts of the painting that followed all agreed on the significance of the obscured canvas for the painting's effects, they disagreed on Foucault's reading of the mirror at the back of the scene. Scrupulously tracing out its perspectival organization, Joel Snyder and Ted Cohen argued in an influential article that, *pace* Foucault, the painting's vanishing point actually coincides with the elbow of the figure standing in the doorway to the right of the mirror portrait. In other words, the images in the mirror can't be those of the live models. Instead, the authors aver, the mirror reflects the painted portrait on the obscured canvas, a claim that had been forwarded at various points in the painting's critical history. In that sense, the authors propose, the painting is meant to affirm the superiority of a self-complete art over nature.[31]

Subsequent analysts have suggested that neither account is fully persuasive: The scale of the depicted canvas isn't right for a portrait painting as that genre was practiced during the era, and in any case Velázquez was always at best approximate in his deployment of perspective.[32] Suggesting that the perspective lines were deliberately left ambiguous, Jonathan Brown, the leading authority on the painter, explains the apparent contradiction by narrativizing it. Noting that the painting was intended for the king's private chambers, he concludes: "If the king were present in person before the picture, he could see, as it were, his own reflection in the mirror. If absent, the picture would be understood as a portrait of the infanta and her retinue, while the mirror-image would be attributed to the reflection from the easel."[33]

Although concurring about the intentional ambiguity of the mirror, for Hubert Damisch such duplicity is not a function of empirical circumstance but is rather an irreducible condition of the image as an image. While the perspectival structure of the Western image has constitutive force vis-à-vis the viewing subject, that formation only becomes a formation by virtue of the torsion or displacement through which it becomes legible as such—to be absolutely absorbed without remainder in the structure would be to be no subject at all. "If there is any representation in painting, the configuration of *Las Meninas* reveals it to consist of a calculated discrepancy between a painting's geometric organization and its imaginary structure,"[34] which is to say, between a symbolic dimension and the identificatory mechanism through which such a structure becomes knowable. We might think of Jacques Lacan's well-known account of Holbein's painting *The Ambassadors*, where the anamorphically distorted image only takes form at the point where the viewer slips out of the painting's perspectival lines, the distorted skull floating in the foreground amounting, according to Lacan, to a figure for the viewer's own fatally eclipsing ensnarement.[35] Insofar as that torsion is constitutive for image and subject alike, representation in the Western tradition inevitably assumes the form it does in Velázquez's opaquely reflexive image: a representation of representation structured around a founding division or visual occlusion.

While for Damisch the aporia of visual figuration is present from the primordial moment Brunelleschi "pierc[ed] a hole in his panel and turn[ed] it around to view it in a mirror," for Lacan, who developed his reflections on *Las Meninas* at the same time Foucault was publishing his and in part in response to the latter, the meaning of the image only comes to view when we factor out reference to a mirror in the painting's construction, such a specular recourse only ever serving to allay the uncanny force of the painting.[36] Rather than depicting its scene as reflected from without in a mirror, the painting with its reversed canvas figures the internal split that constitutes every image as an image. For Lacan, the key to the painting is neither the mirror nor the mystery of what is depicted on the averted canvas but the dehiscence between represented canvas and the frontal plane of the picture itself, the noncoincidence of their representational fields emphasized by the painter's figuring himself as stepping back from and out of the frame of the canvas. All the central figures—the infanta and her attendants, as well as the painter himself—are arrayed, by Lacan's account, within the furrow of that internal divide, an uncanny space governed by the formative divisions and deflections of the gaze.[37] All look out to see what is being depicted before them, and in doing so, all are caught in a gaze from elsewhere, from behind. "You do not see me from where I am looking at you," in Lacan's repeated formulation.[38] Through that rivening of sight, a torsion constitutive of consciousness as well as of world understood as what exceeds and precedes the closed reflex of mirroring reflection, the painting offers up the courtly figures not as subjects or objects of speculative consciousness but in a more primordial condition of sheer display, of spectacle in advance of agential seeing. Through them, the painting figures painting itself not as representation, with all the term implies of transparency, but as what Lacan terms the "representative of representation," the prior, opaque, and unspecularizably material condition that determines the possibility of vision as such.

Whether arrived at via the mirror or by holding the mirror in abeyance, such accounts focused on the divided condition of the look bear out the performative character of the painting—the way it functions as a trap for the gaze. The framing by which *Las Hilanderas* constitutes itself in relation to art historical tradition becomes here, in the context of an explicitly self-representing image, a wild proliferation of frames, the tightening of the spiral of aesthetic reflexive closure yielding only its opposite in a disseminative profusion: Beyond the canvas and the mirror, we have the multiple paintings on the walls, the doorways and windows on the right, the framed doorway opening at the back and its door with its own multiple rectilinear panels. Here too, the painting realizes itself via a logic of redounding visual force. The outright pictorial aggression of the withheld canvas and the painter's direct gaze prompts the viewer to a comparable claim to visual mastery, a mastery measured here as the ability to see the painter's seeing, which is to say, to see beyond the frame of the painter's sight—to see behind the depicted painter. And yet, in staking ourselves on such a posture of exteriority, we find ourselves already inscribed within, captured in the intruding figure peering in from the back of the space from a posture corresponding to the vanishing point of our own sight lines.

Because the image *is* that performative effect, the attempt to objectify it only reduplicates the division by which it reflexively forms: thus the proliferation of delimiting frames, an abyssal continuation suggested by the ascending staircase on which the background visitant stands. The splitting implied by that illimitable framing corresponds with the irreducibility of the painting's status as a scene of interruption. While critics agree that the immediacy of the image relates to the sense that it captures a moment of intrusion, they have notably divided over who is intruding on whom: Has the princess and her entourage happened upon a scene of the painter painting the royal couple? Or have the couple happened upon the scene of Velázquez painting the infanta?[39] For Brown, oddly, the subject of the painting—the king—could be viewed as an intruder not just on the scene but on the painting itself: "The focal point of the picture was the king, who 'interrupted' the figures in *Las Meninas* whenever he entered his summer office."[40] Such ambiguities are irresolvable, for the painting is conjured out of the breach by which the image at once opens and eludes itself, the elusive break of autopoietic formation comparable to the formative intrusions at the heart of *As You Like It*. The indistinction between what arrives from within and what from without, including ourselves, is conveyed by the symmetrical pairing at the back of the painting of the mirror reflecting forward and the doorway opening behind, an equivalency that suggests the grounds of both interior and exterior in the opaque, depthless space of painting as such.

The temporality implied by such a structure becomes apparent when we place pressure on that association between the logic of intrusion and the image's reflexive formation.[41] As Lacan and others have noted, it's no coincidence that the figure who peers in on Velázquez from behind is, in fact, Velázquez: Jose Nieto Velázquez, the king's chamberlain. The painter playfully incorporates at the vanishing point a double as, precisely, the figure who catches him from behind, an apt thematization of the painting's self-incorporative ambitions. That internal disjunction crystallizes in the reflex of the painter's touch. The autoaffective dimension we traced in the spinners here assumes paradigmatic form in the painter's figuration of his own hand. The self-elusive character of an image that originates in its aversive effacement crystallizes in the suspended gesture of a hand that realizes itself as already marked, becomes itself as what already was and will be. In that sense, as formative as Rosalind's tarrying, the painter's interrupted gesture can be seen to embody painting as such, in its irreducible form.[42] As in *Las Hilanderas*, the noncoincidence implied in that reflex is conveyed in the hand's indissociability from techne, here figured as a merging of hand and brush.[43] The implied crossing between human and inhuman is again aligned with animality; the artist's self-inscription—the vanishing point of originary mediation—corresponds with the symmetrically placed dog—a figure for perfect somnolent self-completion.[44] Indeed, we might see in the microdrama of the dwarf's kicking the beast awake a repetition of the painting as scene of artistic coming-to, a playful thematic rendering of the force entailed in that emergence, right down to the prospect of its redounds: Let sleeping dogs lie.

Engaging autopoiesis at the level of the painter's hand means tracing it beyond questions of representation to the relation between sight and tactility as it bears on the possibility of reference as such—the artist's famous "touch" as that which draws to a limit the relation between phenomenal creation and reference. For the fact of *Las Meninas*'s "self-referential" character heightens rather than allays the problem of reference. Insofar as there is no "self"—no object—in advance of the aversive reflex by which the image constitutes itself, the "self-referential" image will be the one that raises the question of the constitutive force of the movement of reference as such, a referring that forms itself precisely by interrupting itself, breaking beyond itself as reflective, meaning forming structure. That that movement should assume, as Redfield suggests, the form of an *insistence*—a disturbance—can be sensed in the way it intrudes itself into the act of theoretical reflection.

For Foucault, the issue of reference arises in relation to his remarks on the importance of not invoking the proper name, which in this case means the name of the king, Philip IV. The name belongs to the dimension of language, and language

> n'est pas celui que déploient les yeux, mais celui que définissent les sucessions de la syntaxe. Or le nom propre, dans ce jeu, n'est qu'un artifice: il permet de montrer du doigt, c'est-à-dire de faire passer subrepticement de l'espace où l'on parle à l'espace où l'on regarde, c'est-à-dire de les refermer commodément l'un sur l'autre comme s'ils étaient adéquats. Mais si on veut maintenir ouvert le rapport du langage et du visible, si on veut parler non pas à l'encontre mais à partir de leur incompatibilité, de manière à rester au plus proche de l'un et de l'autre, alors il faut effacer les noms propres et se maintenir dans l'infini de la tâche.[45]
>
> [is not that deployed by our eyes but that defined by the sequential elements of syntax. And the proper name, in this particular context, is merely an artifice: It gives us a finger to point with, in other words, to pass surreptitiously from the space where one speaks to the space where one looks; in other words, to fold one over the other as though they were equivalents. But if one wishes to keep the relation of language to vision open, if one wishes to treat their incompatibility as a starting point for speech instead of as an obstacle to be avoided, so as to stay as close as possible to both, then one must erase those proper names.]

Foucault's aim here is to head off recourse to the "anecdotal," by which he means narrative history, as an explanatory ground for the image, an image that in this case engages the more fundamental conditions of representation. Still, to cast this concern in terms of reliance on "a finger to point with" is to raise the issue not just of narrative but of reference. For Foucault, pointing implies a gesture that surreptitiously bypasses the distinction between discourse and vision, a false effect insofar as language and vision are by definition distinct domains. And yet, Foucault's claim to the self-evidence of that distinction is belied by the force of the eradication to which he enjoins us: "One must erase those proper names." Repeating the self-cancelling logic of the painting, Foucault curiously evokes the all-but-erased figures in the mirror, "the palest, the most unreal, the most compromised of all the painting's images," as if in refusing to point he nevertheless points there, to the core of the image. Thus the painting's power of capture, but here in a form that suggests an untoward crossover between language—what is, for Foucault, "merely an artifice"—and phenomenal seeing, an impurity in what he takes to be the pure domain of sight.

As if in acknowledgment of this intractability, rather than being erased, reference is "elided" according to Lacan. Referring to the indirections of gaze that expose painting in its character as *objet a*, that is, in its unspecularizable material status, Lacan remarks:

> « Tu ne me vois pas d'où je te regarde »... je n'ai point dit « Tu ne me vois pas, là, d'où je te regarde », que le « là » est élidé, ce « là » sur lequel la pensée moderne a mis tant d'accent sous la forme du dasein comme si tout était résolu, de la fonction de l'être ouvert à ce qu'il y ait un être là. Il n'y a pas de « là » où VELÀZQUEZ, si je le fais parler, invoque dans ce « Tu ne me vois pas d'où je te regarde ». À cette place béante, à cet intervalle non marqué, est précisément ce « là », où se produit la chute de ce qui est en suspens sous le nom de l'objet a.[46]
>
> [I say, "you do not see me from where I am looking at you," not "you do not see me, there, from where I am looking at you"... rather, the "there" is elided, this "there" on which modern thinking has put so much accent in the form of the *dasein*, as if everything was thus determined, from the function of the open being to the fact that there is a being there. There is no "there" that Velázquez, if I may make him speak, invokes, in this "you do not see me from where I am looking at you." In this gaping place, in this "there," where there is produced the fall of what is in suspense under the name of *objet a*.]

There is no "there" in this not seeing from where I look because the "object" in this case consists of nothing but the reiterative effect by which the image at once eludes and captures us. And yet, the "there" recurs, precisely in reference to the elision, the "gap," in which occurs the "fall" of the *objet a*. Or, it does and doesn't recur. Much depends on an equivocation in the meaning of quotation marks in the passage, the uncertain shift from an indicator of mere facticity—the nonexistence of reference—to the bracketed ontology—nonpresent presence—implied by a quotation's suspension of proper meaning. Figuring representation as from the outset a "representative of representation," quotation translates the determining cut or intrusion of the diacritical frame into linguistic terms.

And here too, as in Foucault, that equivocal moment of nonreferential reference is associated with an instance of force involving a crossing between seeing and speaking. "There is no 'there' that Velázquez, if I may make him speak, invokes": It is no coincidence that the uncertain status of quotation arises in the context of a scene of ventriloquizing, a ventriloquizing that amounts to an apostrophe insofar as it involves giving voice to what is by definition mute. Despite his insistence, like Foucault's, on the discreteness of the visual domain, Lacan secures the meaning of the image by speaking for Velázquez, an instance of forced identification in the context of an argument staked against mirroring identification. A case of coercive animation—of kicking the dog to life—the passage suggests a troubling of the distinction between speech and sight as the condition for the phenomenal "thereness" of the image—think of *King Lear*'s letters as suns. Comparable to the force that precedes and determines the distinction between auto- and heteroreference in Luhmann, the apostrophe locates identification not as a consolidating specular function but in relation to a referential drive more primordial than the opposition between seeing and blindness implied by the turned canvas or by any anamorphic distortion of sight lines, a positing that necessarily implicates theoretical thought insofar as it at once enables and destabilizes the space of tropic oppositions that underwrites cognition.[47]

Such instances of theoretical perturbation suggest that we discern the force of reference and thus of history neither in the painting as a mirror of the world nor by setting the mirror aside but as an exorbitancy within the workings of identification itself. There is little doubt that Velázquez consciously deployed the ambiguous potential of an included mirror. Consider the device as playfully held up by the fluttering cupid in the *Rokeby Venus*. Though the mirror shows us the reclining figure's face, according to the sight lines of the image, Jonathan Brown demonstrates, the glass should reflect the figure's genital zone. We can read this as a simple joking ploy: deflecting the viewer's shameful desires toward a more elevated reflection—toward art. But it would be more accurate to say that in such deflection, the image does in fact give us what we want. Just insofar as we catch Venus in her act of solitary self-reflection, we take on the erotically charged posture of the voyeur. Then again, this is Venus, and we can't be certain that she isn't conscious of the onlooker's gaze. With that prospect, the painting assumes the erotic form specific to the image as image, its vacillating status as that which does and does not look back, at once mirror and painting.

Las Meninas realizes a comparable logic as it bears on the relation between sovereign power and aesthetic formation. As we've seen, the painting has occasioned a debate concerning whether the mirror at the back of the space reflects the regal couple as the extrinsic object of the painter's gaze or the surface of the canvas whose back we see at the left. On the one hand, the image figures the painting as sovereign self-reflective totality at the expense of what lies beyond it; on the other, it incorporates that beyond at the cost of an opacity at the painting's center. The referential dimension of the painting corresponds, I would argue, neither to the one possibility nor to the other but to the necessity and impossibility of their convergence, the image emerging only via the splitting that opens the distinction between external and internal, within and beyond, in the first instance. Reference occurs as that insistent though inapprehensible cut, that which constitutes the possibility of phenomenal vision even as it establishes the space of speculative thought.[48]

Locating the sovereign at that point of referential insistence means positing sovereignty as something purely phantasmatic, like what might be conjured from the depths of two mirrors placed face to face: ghostly insofar as it appears at once as effect and cause of the structure of substitution that sets phenomenal meaning in motion. Sovereignty as absolute referent in that sense emerges out of precisely the contingency and historicity sovereignty has always explicitly defined itself against: no coincidence perhaps that Velázquez's art coincides with the decline of empire. That that phantasm is our phantasm—that the mirror-picture does and does not reflect us as viewers as well—suggests the extent to which the ghost image conjures our own fantasy of a sovereign ground, a pure projection that nevertheless precedes us: at once mirror and painting. Reiterating all the other doublings involved in the image—subject and object, viewer and viewed, sovereign and subject—the double portrait—an anomaly in Velázquez's output and during the era[49]—implies less an object than the prior necessity of a relation to relation, an impossible condition whose abyssal nature is renewed in whatever framing cut one undertakes to grant it objectifiable form. As inscribing reflex, the mirror and the painting locate subjective causation neither then nor now but in the ellipsis that opens even as it undoes such possibilities of narrative historicization.

That site of nonphenomenal inscription, the site where history *insists*, is the point where language and seeing aporetically coincide. We noted the merging of hand and brush, but the tightening of the painting's reflexive spiral intensifies in the apparent continuation of the brush head into a thread of white paint, as if the painter were marking, or defacing, his own image. Condensing the conceptual drama of the self-occluding painting, that incising mark suggests the relation between the figures of disability—the dwarves and cognitively impaired—that populate his images and the intimate grounds of his own art. The vanishing point of the image's self-inclusive, self-exceeding logic is comparable to the weaver's thread: It's a material trace that does and doesn't belong to the painting—note the way the luminous line of white against the black vest repeats the strangely luminous frame or beveled cut around the portrait. As a mark on the painting as much as it is one in the painting the line reads as something like an authorial signature—a point of nominating reference. A signature, but a unique one falling at once within and on the painting insofar as it opens that distinction, thus undoing the reassuring opposition between sight and material inscription. Velázquez's signature is his touch, his touch a signature, one that embodies the painting's affective hold as a trace indissociable from the force we will have deployed to make the image apprehensible.

To read such an unplaceable mark in relation to the force of reference is to understand history according to the logic of interruption that the painting

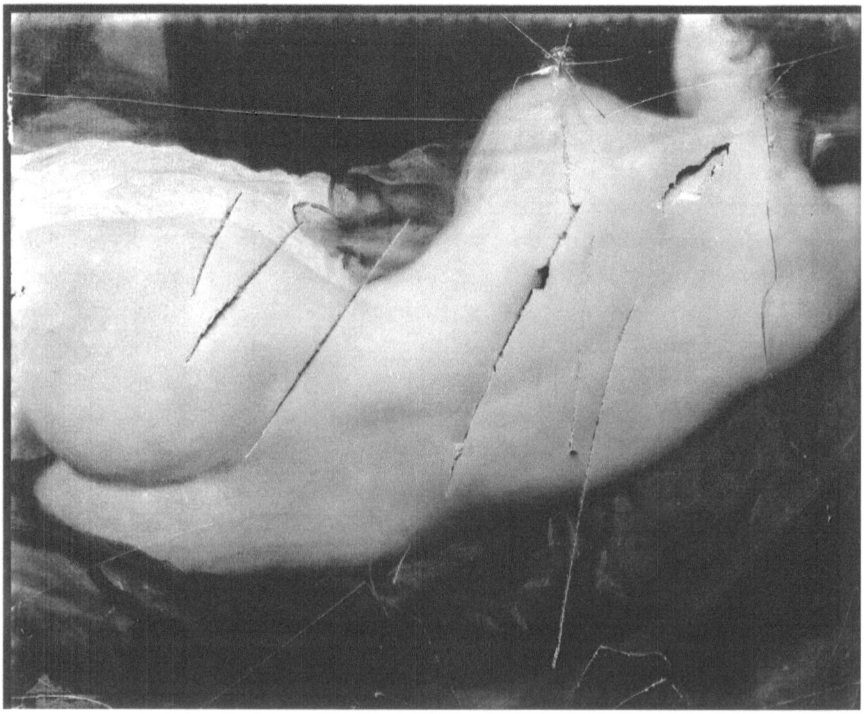

Figure 5: Velázquez, *The Toilet of Venus (Rokeby Venus)*—photo of damage: 1914. Reproduced by permission of the National Gallery, London.

performs—history as that which insistently at once breaches and constitutes itself as cognizable form, as at Actium blinding the eye of phenomenal sight and speculative reason. Given that self-exceeding conception, the historicity as well as the politics of aesthetic formation are perhaps best discerned in the returns of the painting's creative/destructive signature beyond the frame of Velázquez's work and time, the way, for instance, the brush's diagonal incision returns in the slash marks left by the suffragist Mary Robinson on the *Rokeby Venus* in 1914 (see Figure 5). "Justice is an element of beauty as much as colour and outline on canvas," she wrote in defense of her act protesting the imprisonment of a beloved fellow activist.[50] We're back to the domain of Minerva's violative judgment, back to Portia's punitive and saving interventions, and back to the wellsprings of aesthetic form: Robinson's claim isn't for political judgment as against mere beauty but for the violence of justice as beauty's source.

The convergence of artifice, deformation, and the force of justice that crystallize at the site of judgment in the case of Martin Guerre recurs, again at the unstable point of crossover between art and life, the point at which the very distinction between animate and inanimate, human and inhuman remains undecided and where the historicity of the image can't be disentangled from ones disfiguring interpretive intervention or from the way in which, in that very breach, the image renews its power to inscribe. The redounds of that interpretive ensnarement, as well as the political nature of that inscriptive force, are apparent in the contradictions of Robinson's gesture—an act of destruction and of creation, of repudiation and identificatory love, an assertion of sovereign authorial force in the name of a collectivist cause—and in the way such entanglements anticipate Robinson's own fate as a leading figure in the suffragist movement who will also become a leading figure in the movement of British fascism.

Acknowledgments

I want to thank Fordham and Tom Lay, whose tact and intelligence have once again made such a difference. I was unreasonably lucky in my external readers, Russ Leo and Steven Swarbrick, each of whom instantly grasped the stakes of the project and helped point it where it needed to go. Emily Vasiliauskas has had a significant influence on my thinking over the years—I often marvel at my good fortune in winding up with such a colleague. Ted Tregear provided guidance and support from start to finish—his conceptual acuity and readerly gifts have left their mark. Anita Sokolsky read through the entire manuscript more than once and provided her typical nuanced critical insights at important points. I am deeply grateful, too, for John Limon's support and editorial wisdom: I've learned over time simply to do whatever he tells me to do. Chris Warley has been an important interlocutor, though it's been easier to steal his ideas than his flair. I am also indebted to the Renaissance Project and to Colleen Rosenfeld for providing a happy, generative space for ongoing collective thought. I thank Catherine Bates as well for her support and Rob Fellman for his thoughtful editorial care. I shared a version of the comedy chapter with the Harvard Shakespeare Seminar—I'm grateful to fellow participants on that occasion, as well as those at the panel on embodiment at the 2023 Shakespeare convention, including Marjorie Rubright. I'm indebted as well to my coparticipants at a 2024 ACLA seminar on "nuance." Erin Hanson and Anna Swann-Pye, elegant thinkers alike, have each in their own way been an animating resource and a reminder for me of why it is we do what we do. Nothing would have been possible without Renna Pye's love and support. Walter Johnston was an intellectual companion during the time of the book's composition and remains for me a model of uncompromising intellectual commitment—the book is dedicated to his memory.

Notes

Preface

1. Jacques Lacan, *The Four Fundamental Concepts of Psychoanalysis*, ed. Jacques-Alain Miller, trans. Alan Sheridan (Norton, 1977), 60.

2. I want to thank Steven Swarbrick for pointing out the irony of this critical history.

Introduction. The World Elsewhere

1. Jane Hwang Degenhardt and Henry Turner, "Thinking with Cosmology: Aesthetics, Worlds, and Kinds in Shakespeare's *The Tempest*," unpublished manuscript, 21. I want to thank the authors for sharing the essay in advance of its publication in their forthcoming book on the topic of early modern cosmology.

2. Degenhardt and Turner, "Thinking with Cosmology," 3.

3. Degenhardt and Turner, "Thinking with Cosmology," 4.

4. Ayesha Ramachandran, "How to Theorize the World: An Early Modern Manifesto," *New Literary History* 48, no. 4 (2017): 660, 666, 667.

5. Henry Turner, "Worlds of Experience: Fiction, Image, and the Ontology of Ideas in Sidney and Shakespeare," *Studies in English Literature* 62, no. 1 (2022): 234.

6. Turner, "Worlds of Experience," 234.

7. Degenhardt and Turner, "Thinking with Cosmology," 3.

8. Degenhardt and Turner, "Thinking with Cosmology," 3, 7.

9. Degenhardt and Turner, "Thinking with Cosmology," 11.

10. Turner, "Worlds of Experience," 224, 225, 228.

11. Degenhardt and Turner, "Thinking with Cosmology," 17.

12. Degenhardt and Turner, "Thinking with Cosmology," 3.

13. See Marc Redfield, *Phantom Formations: Aesthetic Ideology and the Bildungsroman* (Cornell University Press, 1996), 22–25.

14. Redfield, *Phantom Formations*, 5, 33.

15. Turner, "Worlds of Experience," 228.

16. Bruce R. Smith, *Phenomenal Shakespeare* (Wiley-Blackwell, 2010), 25, 75, 80.

17. Joseph Campana, "On Not Defending Poetry: Spenser, Suffering, and the Energy of Affect," *PMLA* 120, no. 1 (2005): 36, 45.

18. Karen Raber, *Shakespeare and Posthumanist Theory* (Arden, 2018), 17.

19. Steven Mullaney, *The Reformation of Emotions in the Age of Shakespeare* (University of Chicago Press, 2015), 6; Evelyn Tribble and John Sutton, "Cognitive Ecology as a Framework for Shakespearean Studies," in *Shakespeare and Emotion*, ed. Katherine A. Craik (Cambridge University Press, 2020), 101.

20. Brian Massumi, *Parables for the Virtual: Movement, Affect, Sensation* (Duke University Press, 2002), 28.

21. Giulio J. Pertile, *Feeling Faint: Affect and Consciousness in the Renaissance* (Northwestern University Press, 2019), 135.

22. Campana, "On Not Defending Poetry," 46.

23. Jean-Luc Nancy, *A Finite Thinking*, ed. Simon Sparks (Stanford University Press, 2003), 92.

24. On such a *kataskopic* "view from above" in the early modern context, see Ayesha Ramachandran, *The Worldmakers: Global Imagining in Early Modern Europe* (University of Chicago Press, 2018), 112.

25. William Shakespeare, *As You Like It*, ed. Alan Brissenden (Oxford University Press, 1993), 2.7. 163–66. Cited hereafter parenthetically.

26. The world as "at once, other than itself at the very moment of its coming-to-be." Jonathan Goldberg, Karen Newman, and Marcie Frank, "Introduction: World Enough and Time," in *This Distracted Globe: Worldmaking in Early Modern Europe*, ed. Jonathan Goldberg, Karen Newman, and Marcie Frank (Fordham University Press, 2016), 3.

27. Jean-Luc Nancy, *The Creation of the World or Globalization*, trans. François Raffoul and David Pettigrew (State University of New York Press, 2007), 42.

28. On the "pre-existence of the gaze—I see from only one point, but in my existence am looked at from all sides"—see Jacques Lacan, *The Four Fundamental Concepts of Psychoanalysis*, ed. Jacques-Alain Miller, trans. Alan Sheridan (Norton, 1977), 72. The account of the relation between aesthetics and world proposed here should be distinguished from the instrumental and constructivist terms implied by Ramachandran, for whom "the world itself" during early modernity "becomes an aesthetic object that can be made and unmade by individuals," a function of "an ethically motivated vision of the world . . . emerging from the individual subjectivity of a particular human maker." Ramachandran, *Worldmakers*, 224, 227.

29. Niklas Luhmann, *Introduction to Systems Theory*, trans. Peter Gilgen (Polity, 2013), 62–64.

30. Luhmann, *Introduction to Systems Theory*, 44.

31. Martin Heidegger, *Being and Time*, trans. Joan Stambaugh (State University of New York Press, 2010), section 26. On Heidegger's conception of "worlding," see especially Pheng Cheah, *What Is a World? On Postcolonial Literature as World Literature* (Duke University Press, 2016), 95–129. My account of Luhmann's pertinence for a conception of world in the early modern context should be distinguished from David Glimp's perspectival understanding of the implication of Luhmann's account of framing—a matter of opening a secondary posture from which one might acknowledge the limits of a hegemonic culture. David Glimp, "Extreme Cary," in *This Distracted Globe*, ed. Jonathan Goldberg, Karen Newman, and Marcie Frank (Fordham University Press, 2016), 57–59. Rather than offering a different vantage on the world, autopoiesis as I engage it implies the constitution of the possibility of world as such.

32. "When handling a distinction, you always have a blind spot or something invisible behind your back." Luhmann, *Introduction to Systems Theory*, 104.

33. Degenhardt and Turner, "Thinking with Cosmology," 3, 6.

34. Mary Floyd-Wilson and Garrett A. Sullivan Jr., "Introduction: Inhabiting the Body, Inhabiting the World," in *Environment and Embodiment in Early Modern Drama and Performance*, ed. Mary Floyd-Wilson and Garrett A. Sullivan Jr. (Palgrave Macmillan, 2007), 6; and John Sutton, "Spongy Brains," in *Environment and Embodiment*, 15; Amanda Bailey, "Speak What We Feel: Sympathy and Statecraft," in *Affect Theory and Early Modern Texts: Politics, Ecologies, and Form*, ed. Amanda Bailey and Mario DiGangi (Palgrave Macmillan, 2017), 36.

35. Ruth Ronen, *Possible Worlds in Literary Theory* (Cambridge University Press, 2009), 39.

36. Ronen, *Possible Worlds in Literary Theory*, 92.

37. Ronen, *Possible Worlds in Literary Theory*, 95.

38. Ronen, *Possible Worlds in Literary Theory*, 92. On the illimitable or unsaturatable character of context, see Jacques Derrida, "Signature, Event, Context," in *Margins of Philosophy*, trans. Alan Bass (University of Chicago Press, 1982), 307–21.

39. Simon Palfrey, *Shakespeare's Possible Worlds* (Cambridge University Press, 2017), 114.

40. Palfrey, *Shakespeare's Possible Worlds*, 37.

41. Palfrey, *Shakespeare's Possible Worlds*, 124, 126.

42. Palfrey, *Shakespeare's Possible Worlds*, 124.

43. Palfrey, *Shakespeare's Possible Worlds*, 55.

44. Palfrey, *Shakespeare's Possible Worlds*, 121.

45. Palfrey, *Shakespeare's Possible Worlds*, 37, 124.

46. On aesthetics and phenomenality, see Paul de Man, *Aesthetic Ideology*, ed. Andrzej Warminski (University of Minnesota Press, 1966), 61–70.

47. Palfrey, *Shakespeare's Possible Worlds*, 240.

48. Degenhardt and Turner, "Thinking with Cosmology," 33.

49. William Shakespeare, *The Tempest*, ed. Stephen Orgel (Clarendon, 1987), 3.2.135–43. Cited hereafter parenthetically.

50. Degenhardt and Turner, "Thinking with Cosmology," 34.

51. It is this constitutive dimension of the undecidable limits of theater that distinguishes an autopoietic account from the "phenomenology of uncertainty" forwarded in reception-theory accounts such as Lauren Robertson's, which persuasively argues that early modern drama "celebrated the theater's ability to create experiences of pleasurable unknowing." Lauren Robertson, *Entertaining Uncertainty in the Early Modern Theater: Stage Spectacle and Audience Response* (Cambridge University Press, 2023), 14. As Robertson presents them, such representational instabilities are an authorial choice and a reflection of the phenomenological procedures of an already presumed apprehending subject, as well as of the epistemological anxieties prompted by an already presumed world.

52. Niklas Luhmann, *Art as Social System*, trans. Eva M. Knodt (Stanford University Press, 2000), 61.

53. On that separation as an early modern phenomenon, see Luhmann, *Art as Social System*, 252.

54. Claude Lefort, *Democracy and Political Theory*, trans. David Macey (University of Minnesota Press, 1988), 225, 232.

55. On the implicit exorbitancy of the perspectival construct, see Luhmann, *Introduction to Systems Theory*, 113; on the epistemological anxiety provoked by the contingency of perspectival knowledge, see Ramachandran, *The Worldmakers*, chap. 1.

56. Luhmann, *Art as Social System*, 264–65. On the emergence of art as a self-conscious system during the Renaissance, see Robert Williams, "Italian Renaissance Art and the Systematicity of Representation," *Rinascimento* 43 (2004): 309–22.

57. Niklas Luhmann, *Observations on Modernity*, trans. William Whobrey (Stanford University Press, 1998), 51–52. For Luhmann, the advent of a monotheistic notion of a creator god meant, inextricably, an observer god; the constitutive finitude of the world understood as created also means the contingency implicit in the fact of that world's being conceived under a gaze from elsewhere and an understanding of consciousness as the condition of seeing oneself as seen. Such an irreducibly mediated condition of seeing oneself seen is, of course, heightened with Protestantism, with its translation of faith into the always uncertain business of reading the signs of one's salvation.

58. William Shakespeare, *King Lear*, ed. R. A. Foakes (Arden Shakespeare, 1997), 4.6.20–23.

59. Luhmann, *Introduction to Systems Theory*, 51.

60. On the distinction between the determinate reversals of reason and unreason and the more radically unstable determinate undoing of determination implied by the problem of reference, see Paul de Man, "Hypogram and Inscription," in *The Resistance to Theory* (University of Minnesota Press, 1986), 42–43.

61. Walter Benjamin, "Theses on the Philosophy of History," in *Illuminations: Essays and Reflections*, ed. Hannah Arendt (Harcourt, 1968), 254.

62. Jean-Luc Nancy, *After Fukushima: The Equivalence of Disasters*, trans. Charlotte Mandell (Fordham University Press, 2014), 31.

63. Nancy, *After Fukushima*, 18–19.

64. Nancy, *After Fukushima*, 26.

65. The following paragraphs expand a reading forwarded in my *The Storm at Sea: Political Aesthetics in the Time of Shakespeare* (Fordham University Press, 2015), 150–51.

66. While I fully endorse Ramachandran's conviction that the category of world holds the potential to "resist . . . the hegemonic energies of empire," I am not persuaded that the means to that potential can be found in the reflective terms within which she frames it, as a matter of "the mutually constitutive interplay between self and world"—world as resistance and as cause begins, I'm suggesting, where such a mutually reinforcing binary falters. Ramachandran, *Worldmakers*, 77.

67. The temporal paradox of the Renaissance as a category has been critically engaged in an entire range of contexts, including most recently, and with particular nuance, in William West's forthcoming book on the concept's historical afterlives. For an especially subtle and comprehensive account of the idea of the Renaissance as a specifically aesthetic mode whose ongoing promise is inextricable from its historicity, see Christopher Warley, *Auerbach's Renaissance: Rebirths of an Aesthetic from Shakespeare to Ferrante* (Cambridge University Press, 2025).

1. The Autopoietics of Tragedy

1. Craig Dionne, *Posthuman Lear* (Punctum, 2016), 167.

2. Rosie Braidotti, "The Politics of 'Life Itself' and New Ways of Dying," in *New Materialisms: Ontology, Agency, and Politics*, ed. Diana Coole and Samantha Frost (Duke University Press, 2010), 212.

3. Braidotti, "The Politics of 'Life Itself' and New Ways of Dying," 214.

4. Claire Colebrook, *Death of the Posthuman: Essays on Extinction* (Open Humanities Press, 2015), 1:177. See as well Steven Swarbrick's cogent critique of claims made by affirmative variants of vitalism for "life itself," instead "pushing beyond the here and now, the lived, toward an ethics of the unlivable, where life and nonlife con-*volve*." Steven Swarbrick, *The Environmental Unconscious: Ecological Poetics from Spenser to Milton* (University of Minnesota Press, 2023), 229. On the "deep-seated vivophilia that currently underwrites our posthuman moment," see also Steven Swarbrick and Karen Raber, "Introduction: Renaissance Posthumanism and Its Afterlives," *Criticism* 62, no. 3 (2020): 314.

5. Dionne, *Posthuman Lear*, 148, 176.

6. Derrida, "The Ends of Man," in *Margins of Philosophy*, trans. Alan Bass (University of Chicago Press, 1972), 111–36.

7. Karen Raber, *Shakespeare and Posthumanist Theory* (Arden, 2018), 17.

8. Ian Bogost, *Alien Phenomenology, or What It's Like to Be a Thing* (University of Minnesota Press, 2012), 31.

9. Bogost, *Alien Phenomenology*, 51.

10. Bogost, *Alien Phenomenology*, 12.

11. Bogost, *Alien Phenomenology*, 25.

12. Bogost, *Alien Phenomenology*, 58.

13. Cary Wolfe, *What Is Posthumanism?* (University of Minnesota Press, 2009), chap. 1.

14. Niklas Luhmann, *Observations on Modernity* (Stanford University Press, 1998), 19.

15. See Wolfe, *What Is Posthumanism?*, xxiii.

16. Colebrook, *Death of the Posthuman*. For a subtle, psychoanalytically oriented account of the environmental implications of such a formatively divided and noncoinciding conception of world, see Swarbrick, *The Environmental Unconscious*, 16–53.

17. William Shakespeare, *Othello*, ed. Michael Neill (Oxford University Press, 2008), 2.3.331–33, 3.3.369, 364. On tragedy as redounding snare, see Peter Szondi, *An Essay on the Tragic* (Stanford University Press, 2002), 59, 67–68.

18. William Shakespeare, *Anthony and Cleopatra*, ed. Michael Neil (Oxford University Press, 1994), 4.9.17–18.

19. On the latter, see Philippe Lacoue-Labarthe, "Caesura of the Speculative," in *Typography: Mimesis, Philosophy, Politics*, trans. Christopher Fynsk (Harvard University Press, 1989), 208–35.

20. Joseph Campana and Scott Maisano, "Introduction," in *Renaissance Posthumanism*, ed. Joseph Campana and Scott Maisano (Fordham University Press, 2016), 13.

21. Though a possibility perhaps already implied by the fact that "hamartia," the tragic flaw as Aristotle describes it, literally means "to miss the mark." See Dennis J. Schmidt, *On Germans and Other Greeks: Tragedy and Ethical Life* (Indiana University Press, 2001), 281.

22. See Thomas M. Greene, *The Light in Troy: Imitation and Discovery in Renaissance Poetry* (Yale University Press, 1986).

23. "The tragic marks a boundary of the realm of art no less than the field of history." Walter Benjamin, "Trauerspiel and Tragedy," in *The Origin of the German Trauerspiel*, trans. Howard Eiland (Harvard University Press, 2019), 261.

24. Stephen Greenblatt, "Psychoanalysis and Renaissance Culture," in *Learning to Curse: Essays in Early Modern Culture* (Routledge, 1990), 184.

25. Bruno Latour, *Reassembling the Social: An Introduction to Actor-Network Theory* (Oxford University Press, 2007), 133.

26. Greenblatt, "Psychoanalysis and Renaissance Culture," 184, 186–87.

27. Greenblatt, "Psychoanalysis and Renaissance Culture," 187.

28. Greenblatt, "Psychoanalysis and Renaissance Culture," 188.

29. Natalie Z. Davis, *The Return of Martin Guerre* (Harvard University Press, 1983), 21, 34. Subsequent references will be cited parenthetically.

30. Focusing on the interim of Hamlet's sea voyage, Meredith Evans writes acutely about the gap as that which constitutes the possibility of narrative representation. Meredith Evans, "Cities of the Stranger," in *This Distracted Globe: Worldmaking in Early Modern Europe*, ed. Jonathan Goldberg, Karen Newman, and Marcie Frank (Fordham University Press, 2016), 148.

31. Jacques Lacan, *The Four Fundamental Concepts of Psychoanalysis*, ed. Jacques-Alain Miller, trans. Alan Sheridan (Norton, 1977), 232.

32. At another point, Davis explicitly enters into a "thought experiment" in which she explains Arnaud's unaccountable knowledge by imagining a moment of encounter between the real husband in his wanderings and the pretender. They meet, recognize the uncanny resemblance, and "exchange confidence": "Martin expresses himself with ambivalence about his patrimony and his wife, perhaps seems to imply to his look-alike, 'take her.' And Pansette says to himself, 'Why not?'" (39). Against the prospect of limitless mediation—no end of stand-ins—Davis conceives a moment of epistemological transparency by translating Bertrande into the explicit object of exchange between two knowing men, indulging at least at the level of fantasy in a form of trafficking in women far more outright than the story's more diffuse versions of that economy.

33. Aristotle, *The Poetics*, trans. S. H. Butcher (Cosimo, 2008), sec. IX.

34. Kathy Eden, *Poetic and Legal Fiction in the Aristotelian Tradition* (Princeton University Press, 1986), 25.

35. Lorna Hutson, *The Invention of Suspicion: Law and Mimesis in Shakespeare and Renaissance Drama* (Oxford University Press, 2007), 3–30.

36. Eden, *Poetic and Legal Fiction in the Aristotelian Tradition*, 91, 92. For an especially refined analysis of the religious stakes of tragedy and questions of causality in the era, see Russ Leo, *Tragedy as Philosophy in the Reformation World* (Oxford University Press, 2019).

37. Schmidt, *On Germans and Other Greeks*, 99. See G. W. F. Hegel, *The Phenomenology of Spirit*, trans. Terry Pinkard (Cambridge University Press, 2018), 257–61.

38. Hegel, *The Phenomenology of Spirit*, 271–73. See Joshua Billings, *Genealogy of the Tragic: Greek Tragedy and German Philosophy* (Princeton University Press, 2014), 173–82.

39. Kathleen M. Sands, "Tragedy, Theology, and Feminism in the Time After Time," in *Rethinking Tragedy*, ed. Rita Felski (Johns Hopkins University Press, 2008), 82.

40. Eden, *Poetic and Legal Fiction in the Aristotelian Tradition*, 22.

41. Aristotle, *Poetics*, sec. IX.

42. "The meaning of fulfilled time in the tragic fate emerges in great moments of passivity: in the tragic decision, in the retarding moment, in catastrophe." Benjamin, "Trauerspiel und Tragodie," 263. On the relation between passion and passivity,

see Jacques Derrida, *The Animal That Therefore I Am*, trans. David Wills (Fordham University Press, 2008), 396.

43. On the self-exceeding structure of aestheticization, see Paul de Man, *Aesthetic Ideology* (University of Minnesota Press, 2003), 61–69; Marc Redfield, *The Politics of Aesthetics: Nationalism, Gender, Romanticism* (Stanford University Press, 2003), 10; Christopher Pye, *The Storm at Sea: Political Aesthetics in the Time of Shakespeare* (Fordham University Press, 2015), 14–16.

44. See Redfield, *Politics of Aesthetics*, 112–20; and Jean-Luc Nancy, "Res ipsa et ultima," trans. Steven Miller, in *A Finite Thinking*, ed. Simon Sparks (Stanford University Press, 2003), 314.

45. Michel de Montaigne, *The Complete Essays of Montaigne*, trans. Donald Frame (Stanford University Press, 1976), 288, 248.

46. On the development of that thought, see Richard Scholar, *The* Je-Ne-Sais-Quoi *in Early Modern Europe* (Oxford University Press, 2005).

47. Montaigne, *Essays*, 285.

48. Montaigne, *Essays*, 788.

49. Montaigne, *Essays*, 791.

50. Montaigne, *Essays*, 791.

51. Jean-Luc Nancy, *The Creation of the World or Globalization*, trans. François Raffoul and David Pettigrew (State University of New York Press, 2007), 100. See Christopher Pye, "'No Cause, No Cause': *King Lear* and the Space of the World," in *Political Aesthetics in the Era of Shakespeare*, ed. Christopher Pye (Northwestern University Press, 2020), 47. I reiterate here a number of the claims formulated in that essay.

52. William Shakespeare, *King Lear*, ed. R. A. Foakes (Arden Shakespeare, 1997), 1.2.99–100. Subsequent references will be to this edition and cited parenthetically.

53. The following analysis draws from Christopher Pye, "Green World / No World: *King Lear*, Eco-Criticism, and the Politics of Finitude," *Shakespeare Jahrbuch* 153 (2017): 11–34.

54. Annabel Patterson celebrates it as "Lear's great speech on social inversion." Annabel Patterson, *Shakespeare and the Popular Voice* (Blackwell, 1989), 112.

55. On nearness—actual proximity—as that which preserves the far as its condition, see Martin Heidegger, "The Thing," in *Poetry, Language, Thought*, trans. Albert Hofstadter (HarperCollins, 1971), 177–78.

56. On a noninstrumentalized techne as that which "gives man entry into something which, of himself, he can neither invent nor in any way make," see Martin Heidegger, "The Question Concerning Technology," in *Basic Writings*, trans. David Krell (HarperCollins, 2008), 313.

57. Jean-Luc Nancy, *After Fukushima: The Equivalence of Catastrophes*, trans Charlotte Mandell (Fordham University Press, 2015), 12.

58. Nancy, *After Fukushima*, 25–26.

59. Mary Floyd-Wilson and Garrett Sullivan Jr., "Introduction: Inhabiting the Body, Inhabiting the World," in *Environment and Embodiment in Early Modern*

England," ed. Mary Floyd-Wilson and Garrett Sullivan Jr. (Palgrave, 2007); and John Sutton, "Spongy Brains," in *Environment and Embodiment in Early Modern England*, 6, 15.

60. The distinction between such a reading of the play's relation to world and cognitive-theory approaches should be evident. Even with "Distributed Cognition" and "Extended Mind" models, where cognition is taken to be fluidly "distributed across body, brain and world," those categories—"brain" and "world," "insides, objects and people"—are empirically presupposed. At stake in the drama is the question of how those distinctions are posited in the first place. See Evelyn Tribble, *Cognition in the Globe: Attention and Memory in Shakespeare's Theatre* (Palgrave, 2011), 6–7, 12.

61. On the Heideggerian themes here—home as what is opened by the breach of an uncanny homelessness; man's untimely being as an effect of that breach and thus human existence as irreducibly historical event or "in-cident" rather than what is "present at hand"; the role of techne in that opening onto historical Dasein—see Martin Heidegger, *Introduction to Metaphysics*, trans. Gregory Fried and Richard Polt (Yale University Press, 2000), 174, 178–79. On relation as what appears where world as such cannot be presupposed, see Jacques Derrida, *The Beast and the Sovereign* (University of Chicago Press, 2011), 2:221; and Jacques Derrida, "Rams: Uninterrupted Dialogue—between Two Infinities, the Poem," in *Sovereignties in Question: The Poetics of Paul Celan*, ed. Thomas Dutoit and Outi Pasanen (Fordham University Press, 2005), 149–53.

2. *Timon of Athens*, Autopoiesis, and the Political-Aesthetic Foundations of the Polis

1. William Shakespeare and Thomas Middleton, *Timon of Athens*, ed. John Jowett (Oxford University Press, 2004), 12.1–6. Subsequent references to the play will be to this edition and cited parenthetically.

2. Giorgio Agamben, *Homo Sacer: Sovereign Power and Bare Life*, trans. Daniel Heller-Roazen (Stanford University Press, 1998), 85.

3. Claude Lefort, "The Permanence of the Theological-Political," in *Democracy and Political Theory*, trans. David Macey (University of Minnesota Press, 1988), 225. On the distinction between oppositional conflicts within the social and the objectless force that constitutes the social as such, see Ernesto Laclau, *The Rhetorical Foundations of Society* (Verso, 2014), 30–32.

4. On the friend-enemy distinction as foundational, see Carl Schmitt, *The Nomos of the Earth in the International Law of Jus Publicum Europaeum*, trans. G. L. Ulmen (Telos, 2006), 67–79.

5. See Agamben, *Homo Sacer*, 104–11.

6. Coppelia Kahn, "'Magic of Bounty': *Timon of Athens*, Jacobean Patronage, and Maternal Power," *Shakespeare Quarterly* 38, no. 1 (1987): 34–57.

7. Ken Jackson, "One Wish or the Possibility of the Impossible: Derrida, the Gift,

and God in *Timon of Athens*," *Shakespeare Quarterly* 52, no. 1 (2001): 34–66. See also Julia Lupton, *Thinking with Shakespeare* (University of Chicago Press, 2011), 146.

8. Jackson, "One Wish," 47, 59.

9. Luke Wilson, "Promissory Performances," *Renaissance Drama* 25 (1994): 79.

10. Michel de Montaigne, *The Complete Essays of Montaigne*, ed. Donald Frame (Stanford University Press, 1965), 79.

11. Montaigne, *The Complete Essays of Montaigne*, 141.

12. Montaigne, *The Complete Essays of Montaigne*, 139.

13. Laura Kolb, "Debt's Poetry in *Timon of Athens*," *Studies in English Literature* 58, no. 2 (2018): 45.

14. Lesley Brill, "Truth and *Timon of Athens*," *Modern Language Quarterly* 40, no. 1 (1979): 22.

15. "Savor [my body parts] well," the cannibal's victim declares, "you will find in them the taste of your own flesh." Montaigne, *The Complete Essays of Montaigne*, 158.

16. See Schmitt, *Nomos*, 67–79.

17. Agamben, *Homo Sacer*, 29.

18. Agamben, *Homo Sacer*, 107.

19. See Agamben, *Homo Sacer*, 26. The instrumentalizing language of capture recurs at such moments throughout Agamben's oeuvre; see Giorgio Agamben, *The Kingdom and the Glory*, trans. Lorenzo Chiesa (Stanford University Press, 2011), 251; and Giorgio Agamben, *The Use of Bodies*, trans. Adam Kotsko (Stanford University Press, 2016), 237, 265.

20. See Agamben, *Homo Sacer*, 51.

21. For an especially acute account of aesthetics as the reflexively divisive operation through which it constitutes its own ground, see Marc Redfield, *The Politics of Aesthetics: Nationalism, Gender, Romanticism* (Stanford University Press, 2003).

22. Giorgio Agamben, *The Man Without Content*, trans. Georgia Albert (Stanford University Press, 1999), 72.

23. See Agamben, *The Man Without Content*, 36.

24. Agamben, *The Man Without Content*, 37, 100, 102.

25. In conceiving aesthetics in relation to the conditions of phenomenal experience, the play engages the category in the fully Kantian sense. See Paul de Man, *Aesthetic Ideology* (University of Minnesota Press, 1996), 61–69; Redfield, *The Politics of Aesthetics*, 24; Christopher Pye, "Introduction," in *Political Aesthetics in the Era of Shakespeare*, ed. Christopher Pye (Northwestern University Press, 2020); Charles Shepherdson, "Aesthetic 'Sense' in Kant and Nancy," *New Literary History* 48 (2017): 207, 217. Along rather different lines, Joan Pong Linton argues that hunger in the play returns us to use value before the depredations of exchange. Joan Pong Linton, "Timon's Hunger in the Forest: Toward a Political Aesthetics of Being Beside Oneself," in *Political Aesthetics in the Era of Shakespeare*, 130–35.

26. On the at once autoaffective and other-directed character of Kantian aesthetic sense, see Shepherdson, "Aesthetic 'Sense' in Kant and Nancy," 210, 212.

27. Jacques Rancière, *Aesthetics and Its Discontents* (Polity, 2009), 8.

28. Paul de Man, "Autobiography as De-Facement," in *The Rhetoric of Romanticism* (Columbia University Press, 1984), 46.

29. On catechresis as the limit-point of the social-symbolic, see Ernesto Laclau, *The Rhetorical Foundations of Society* (Verso, 2014), 64.

30. Brent Dawson speaks of Sodom as the point of indistinction between "the fruitfully reproductive slime of the tower and the sterile, perverse gold of the cave" in Spenser's *Faerie Queene*. Brent Dawson, "Worldly Muck: Translating Matter in Book 2 of *The Faerie Queene*," in *This Distracted Globe*, ed. Jonathan Goldberg, Karen Newman, and Marcie Frank (Fordham University Press, 2016), 26.

31. Karl Marx, *Capital*, trans. Eden Paul and Cedar Paul (Dent, 1939), 112–13.

32. On the commodity as "sensuously supersensible"—"a thing without phenomenon," see Jacques Derrida, *Specters of Marx: The State of the Debt, the Work of Mourning, and the New International*, trans. Peggy Kamuf (Routledge, 1994), 150.

33. Marx, *Capital*, 112.

34. "Mammon" means "that which is hidden." Jowett, "Introduction," 62.

35. Robert Darcy, "Shakespeare's Empty Plot: The Epicenotaph in *Timon of Athens*," *Renaissance Drama* 33 (2004): 165, 168–70.

3. A New Air

1. Jane Hwang Degenhardt, "Counter-Imperial Horizons in *Antony and Cleopatra*," *Studies in English Literature 1500–1900* 62, no. 1 (2022): 51.

2. Degenhardt, "Counter-Imperial Horizons in *Antony and Cleopatra*," 66.

3. Degenhardt, "Counter-Imperial Horizons in *Antony and Cleopatra*," 65.

4. On the temporal blurring prompted by the theatricality of the play's ending, see Drew Daniel, *Joy of the Worm: Suicide and Pleasure in Early Modern English Literature* (University of Chicago Press, 2022), 76.

5. William Shakespeare, *Anthony and Cleopatra*, ed. Michael Neill (Oxford University Press, 1994), 4.9.17–18. Subsequent references to the play will be to this edition and cited parenthetically.

6. On the correspondence between Caesar's appropriative triumph and the play itself, see Brent Dawson, "Multiple Worlds in *Antony and Cleopatra*," *Renaissance Drama* 43 (2015): 189.

7. See, for example, Luce Irigaray, *This Sex Which Is Not One*, trans. Catherine Porter (Cornell University Press, 1983), 23–34.

8. On the characters' self-memorializing impulses, see Garrett A. Sullivan Jr., *Memory and Forgetting in English Renaissance Drama: Shakespeare, Marlowe, Webster* (Cambridge University Press, 2005), 104.

9. Janet Adelman, *The Common Liar: An Essay on 'Antony and Cleopatra'* (Yale University Press, 1973), 31–34.

10. On the possibility of representation that is at once substitutive and singular in the play, see Nigel Mapp, "Early Modern Aesthetics: *Antony and Cleopatra* and the Afterlife of Domination," *Aisthesis* 13, no. 2 (2020): 177.

11. On the queen's virtual absence from the scene, see Ania Loomba, *Shakespeare, Race, and Colonialism* (Oxford University Press, 2002), 261. On the persistently narcissistic character of the play's (male) heteroerotic relations, see Jonathan Gil Harris, "'Narcissus in Thy Face': Roman Desire and the Difference It Fakes in *Antony and Cleopatra*," *Shakespeare Quarterly* 45, no. 4 (1994): 408–25.

12. On the recursiveness of "o'erpicturing," see Emily Vasiliauskas, "*Antony and Cleopatra* and the Art of Dying," *Studies in Philology* 117, no. 3 (2020): 511.

13. John Gillies, *Shakespeare and the Geography of Difference* (Cambridge University Press, 1994), 121.

14. William Shakespeare, *The Tempest*, ed. Stephen Orgel (Clarendon, 1987), 3.2.134.

15. On Hercules's presence only in his departing, see Adelman, *The Common Liar*, 80.

16. Including the movement from mimesis toward allegory: Think of the oft-remarked figure of Lamord in *Hamlet*.

17. Alain Badiou, *Being and Event*, trans. Oliver Feltham (Continuum, 2005), 433.

18. See especially William Junker, "The Image of Both Theaters: Empire and Revelation in Shakespeare's *Antony and Cleopatra*," *Shakespeare Quarterly* 66, no. 2 (Summer 2015): 167–87.

19. Adelman, *The Common Liar*, 70–71.

20. David Quint, *Epic and Empire: Politics and Generic Form from Virgil to Milton* (Princeton University Press, 1993), 32.

21. Adelman, *The Common Liar*, 33.

22. It is worth noting the historical distinction between tragic guilt—a classical conception—and tragic shame as Shakespeare engages it here. If the former represents the overdetermined character of the tragic hero's affective response—guilt as a condition that exceeds him—as the intimate measure of a supervening fate, the latter registers that affective overdetermination as a function of the subject's sheerly relational formation, absent transcendental reference. In one regard, the distinction bears out Luhmann's claim for self-conscious observation as a constitutive condition for modernity. In another, though, the play resists such an account, positing affect as the point where such speculative constructions falter, prior to the opposition between subject and object, seer and seen, and situates the fact of historical contingency there.

23. On the experience of return as what initiates historical time, see François Hartog, *Regimes of Historicity: Presentism and Experiences of Time*, trans. Saskia Brown (Columbia University Press, 2003), 54–55, 81, 88.

24. Quint, *Epic and Empire*, 30. As Quint points out, the apotheosis of such a structure occurs with Milton's battle in heaven, where the static equipoise of contesting angelic forces is broken with Christ's intervention, an intervention that can be seen to set worldly history in motion (39).

25. Philippe Lacoue-Labarthe, "Caesura of the Speculative," in *Typography: Mimesis, Philosophy, Politics*, trans. Christopher Fynsk (Harvard University Press, 1989), 45. On reference and the limits of language as a tropic system, see Paul de Man, *Aesthetic Ideology* (University of Minnesota Press, 1993), 61–69. On the origin of nation in the "interruption of teleological and naturalized representations of history," see Meredith Evans, "Cities of Strangers," in *This Distracted Globe: Worldmaking in Early Modern Europe*, ed. Jonathan Goldberg, Karen Newman, and Marcie Frank (Fordham University Press, 2106), 148. Inseparable from the event as such, the reiterative division enacted by the interventional decision amounts, in Badiou's account, to the institution of temporalization. Badiou, *Being and Event*, 210.

26. Virgil, *Aeneid*, trans. A. S. Kline (2002) 8.706–7, https://www.poetryintranslation.com/Admin/Copyright.php.

27. Torquato Tasso, *The Liberation of Jerusalem*, trans. Max Wickert (Oxford University Press, 2009), 16.5–6.

28. Martin Heidegger, *Being and Time*, trans. Joan Stambaugh (State University of New York Press, 2010), 252. On the failure of substitutability as the condition of reference, see Philippe Lacoue-Labarthe, *Poetics of History: Rousseau and the Theater of Originary Mimesis* (Fordham University Press, 2019), 36.

29. Junker, "The Image of Both Theaters," 177–87; Vasiliauskas, "*Antony and Cleopatra* and the Art of Dying," 490–92, 511.

30. The claim I've been forwarding for the irreducibility of such missed encounters should make evident my misgivings about Paul A. Kottman's claims for dialectical "self-realization" in Shakespeare, the argument that *Romeo and Juliet* "is the story of two individuals who actively claim their separate individuality, their own freedom, in the only way they can—through one another." Paul A. Kottman, "Defying the Stars: Tragic Love as the Struggle for Freedom in *Romeo and Juliet*," *Shakespeare Quarterly* 63, no.1 (2012): 6.

31. "How is history apriori possible? When the soothsayer himself shapes and forms the events that he had predicted in advance." Immanuel Kant, *Der Streit der Fakultaten*, in *Kant's Werke: Akademie Textausgabe* 7 (Gruyter, 1968), 79–80. Cited in Reinhart Koselleck, *Futures Past: On the Semantics of Historical Time*, trans. Keith Tribe (Columbia University Press, 2004), 39. And when he does so against his own wishes, we should add.

32. Heidegger, *Being and Time*, 311.

33. Walter Benjamin, "Theses on the Philosophy of History," in *Illuminations*, ed. Hannah Arendt (Schocken, 1968), 255, 261.

34. Junker, "The Image of Both Theaters," 181, 182.

35. Werner Hamacher, "'Now': Walter Benjamin on Historical Time," in *The*

Moment: Time and Rupture in Modern Thought, ed. Heidrun Friese (Liverpool University Press, 2001), 175.

36. Andrew Fichter notes the mistimed, anticipatory, and belated character of Anthony's suicide; Garrett Sullivan observes that the entirety of the play's fifth act occurs within the deferred space of Cleopatra's. Andrew Fichter, "*Antony and Cleopatra*: 'The Time of Universal Peace,'" *Shakespeare Survey* 33 (1980): 108; Sullivan, *Memory and Forgetting*, 103.

37. "I-am-as-having-been": Heidegger, *Being and Time*, 313. On the irreducibility of such a linguistic crossing between animate and inanimate, see Paul de Man, "Autobiography as De-Facement," in *The Rhetoric of Romanticism* (Columbia University Press, 1984), 46.

38. Vasiliauskas, "*Antony and Cleopatra* and the Art of Dying," 518.

39. Catherine Belsey, "Cleopatra's Seduction," in *Alternative Shakespeares 2*, ed. Terence Hawkes (Routledge, 1996), 47. On the scene's failure of representation as suggesting other worlds beyond, see Dawson, "Multiple Worlds," 189.

40. My argument should be distinguished from Lukas Lammer's that Shakespearean drama establishes a recognition of historical relativity based on what he takes to be the "intrinsic doubleness of theatrical communication," its making the past present as what is past. Lukas Lammer, *Shakespearean Temporalities: History on the Early Modern Stage* (Routledge, 2018), 194, *passim*. Such doubleness is neither intrinsic nor stable, *Antony and Cleopatra* suggests—indeed, historicization begins where such a mimetic ratio between past and present falters. This is also to say that Shakespearean history isn't relativistic—a subjectivist claim hinging on point of view—but radically dialectical and thus constitutive of subjectivities.

41. Koselleck, *Futures Past*, 31–33.

42. Hartog, *Regimes of Historicity*, 65–72.

43. Koselleck, *Futures Past*, 104, 34.

44. On the relation between tragedy and the grounds of speculative philosophy, see Suzanne Gearhart, *The Interrupted Dialectic: Philosophy, Psychoanalysis, and Their Tragic Others* (Johns Hopkins University Press, 1992), 160.

45. "Presupposing" not as an ideational proposition but as the precondition of thought. See Heidegger, *Being and Time*, 301.

46. Giorgio Agamben, *The Time That Remains: A Commentary on the Letter to the Romans*, trans. Patricia Dailey (Stanford University Press, 2005), 145.

47. Hamacher, "'Now,'" 185. On temporalization as *ekstasis*, see Heidegger, *Being and Time*, 314–15.

48. "From the first words of the play, our reactions have been at issue." Adelman, *The Common Liar*, 102. On the convergence of nature and fancy in the play, see Ronald R. MacDonald, "Playing Til Doomsday: Interpreting *Antony and Cleopatra*," *Studies in the English Renaissance* 15, no. 1 (1985): 78–90. On being as "being-with," see Jean-Luc Nancy, *A Finite Thinking* (Stanford University Press, 2003), 9: "Groundlessness isn't a lack on the part of being that needs to be undergone, justified, originated. Rather, it is being's reference to nothing, either to substance,

or to subject, not even to being, unless it be *to* a being-to, to itself, to the world as the openness, the throw or the being-thrown of *existence*." See also Jean-Luc Nancy, *Being Singular Plural* (Stanford University Press, 2000), 7–14.

49. Walter Benjamin, "One Way Street," in *Selected Writings, vol. 1: 1913–1926*, ed. Marcus Bullock and Michael W. Jennings (Harvard University Press, 1996), 484. I thank Ted Tregear for pointing out the passage, as well as for his insights concerning the Dolabella exchange.

50. Benjamin, *Illuminations*, 255.

51. Hamacher, "'Now,'" 163.

4. Comedy and Romance

1. Simon Palfrey, *Late Shakespeare: A New World of Words* (Oxford University Press, 1997), 17.

2. Frederick Jameson, "Magical Narratives: On the Dialectical Use of Genre Criticism," in *The Political Unconscious: Narrative as a Socially Symbolic Act* (Routledge, 1981), 89–130.

3. Genre involves not the "context, outside or prehistory of the work, but the very being of the work in time," Julia Lupton acutely observes. Julia Lupton, *Afterlives of the Saints: Hagiography, Typology, and Renaissance Literature* (Stanford University Press, 1996), 38.

4. Steven Mullaney, *The Reformation of Emotions in the Age of Shakespeare* (University of Chicago Press, 2015), 6.

5. Northrop Frye, *The Anatomy of Criticism: Four Essays* (Princeton University Press, 2020); C. L. Barber, *Shakespeare's Festive Comedies* (Princeton University Press, 2012), 12.

6. Northrop Frye, *A Natural Perspective: The Development of Shakespearean Comedy and Romance* (Columbia University Press, 1965), 126.

7. William Shakespeare, *A Midsummer Night's Dream*, Norton Shakespeare, 3rd ed., ed. Stephen Greenblatt et al. (Norton, 2016), 1.1.136, 250–51. Subsequent references to Shakespeare's plays will be to this edition and cited parenthetically.

8. Jacques Lacan, *The Four Fundamental Concepts of Psychoanalysis*, ed. Jacques-Alain Miller, trans. Alan Sheridan (Norton, 1977), 235–36, 251–52.

9. On the contradiction, see Samuel Weber, "Drawing—the Single Trait: Toward a Politics of Singularity," in *Crediting God: Sovereignty and Religion in the Golden Age of Capitalism*, ed. Miguel Vatter (Fordham University Press, 2011), 230. The tension would have been felt in the early modern context in part via the ascendance of *assumpsit* in debt law, with its simultaneous privileging of strict contractualism and of subjective intention—twin features of the modern legal landscape. See Charles Spinosa, "The Transformation of Intentionality: Debt and Contract in *The Merchant of Venice*," *English Literary Renaissance* 24, no. 2 (1994): 372–85; Luke Wilson, "Promissory Performances," *Renaissance Drama* 25 (1994): 67–69; Lynn

Johnson, "Friendship, Coercion, and Interest: Debating the Foundation of Justice in Early Modern England," *Journal of Early Modern History* 8 (2004): 60–64.

10. On Portia robbing Antonio of martyrdom, see Janet Adelman, *Blood Relations: Christian and Jew in 'The Merchant of Venice'* (University of Chicago Press, 2008), 133; and Harry Berger, "Mercifixion in *The Merchant of Venice*," *Renaissance Drama* 38, no. 1 (2010): 159.

11. Here and in what follows I am indebted to Jacques Derrida's account of law as eventless event constituted by its own difference and deferral. See Jacques Derrida, "Before the Law," in *Acts of Literature*, ed. Derek Attridge (Routledge, 1992), 210; Jacques Derrida, "Force of Law: 'The Mystical Foundation of Authority,'" in *Deconstruction and the Possibility of Justice*, ed. Drucilla Cornell, Michel Rosenfeld, and David Gray Carlson (Routledge, 1992), 35–37.

12. On the contradictions of the aesthetic state, see Marc Redfield, *The Politics of Aesthetics: Nationalism, Gender, Romanticism* (Stanford University Press, 2003), 1–44.

13. See Mary Ellen Lamb and Valerie Wayne, "Introduction: Into the Forest," in *Staging Early Modern Romance: Prose Fiction, Dramatic Romance, and Shakespeare*, ed. Mary Ellen Lamb and Valerie Wayne (Routledge, 2009).

14. Palfrey, *Late Shakespeare*, 60.

15. On the "sheen of romance" deriving from the overlaying of the time of the event and the time of its re-cognition, see Lauren Shohet, *Temporality, Genre, and Experience in the Age of Shakespeare: Forms of Time* (Bloomsbury, 2018), 108. On the scene's retrospective/anticipatory status and its implicative effects, see J. K. Barret, "The Crowd in Imogen's Bedroom: Allusion and Ethics in *Cymbeline*," *Shakespeare Quarterly* 66, no. 4 (2015): 440–62. Wendy Beth Hyman provides an especially subtle account of the reflexively involuted feeling of the scene, the way in which Imogen can be seen to be effectively "dreaming" Iachimo's presence, even as he seems to conjure her from his Petrarchan blazons. Wendy Beth Hyman, "'Beyond Beyond': *Cymbeline*, the Camera Obscura, and the Ontology of Elsewhere," *English Literary Renaissance* 52, no. 3 (2022): 409.

16. On such an account of touch, see Jacques Derrida, *On Touching—Jean-Luc Nancy*, trans. Christine Irizarry (Stanford University Press, 2005), 31–38.

17. On language being at once formatively reflexive and tactile in the play, see Palfrey, *Late Shakespeare*, 161.

18. Jodi Mikalachki, "The Masculine Romance of Roman Britain: Cymbeline and Early Modern English Nationalism," in *Shakespeare's Romances*, ed. Alison Thorne (Palgrave, 2003), 118.

19. On the emergence of territory as a relationally defined political category during the era, see Stuart Elden, *Shakespearean Territories* (University of Chicago Press, 2018), 1–8, 26–28.

20. On Milford Haven's liminal status, as well as Britain's divided and singular condition vis-à-vis Wales: Terence Hawkes, *Shakespeare in the Present* (Routledge, 2002), 49–51; Garret A. Sullivan Jr., *The Drama of Landscape: Land, Property, and*

Social Relations in the Early Modern Age (Stanford University Press, 1999), 136–39; Willy Maley, "Cymbeline, the Font of History, and the Matter of Britain," in *Alternative Shakespeares* 3, ed. Diana Henderson (Routledge, 2008), 127–34.

21. "The nation is an ideal community, that is, by definition, either proleptic or passing, ever just beyond reach." Claire McEachern, *The Politics of English Nationhood: 1590–1612* (Cambridge University Press, 1996), 6.

22. Hyman subtly analyzes the moment in relation to the fluidity, reversibility, and, ultimately, the evocative incoherence of perspectivally ordered knowledge in the play. Hyman, "Beyond Beyond," 403–4.

23. Claude Lefort, *Democracy and Political Theory*, trans. David Macey (University of Minnesota Press, 1988), 225, 232.

24. On exposure as the primordial condition of being, see Roberto Esposito, *Communitas: The Origin and Destiny of Community*, trans. Timothy Cambell (Stanford University Press, 2010), 7–12; and Jean-Luc Nancy, *The Creation of the World, or Globalization*, trans. François Raffoul and David Pettigrew (SUNY Press, 2007), 99–101.

25. "This is our responsibility: it isn't a task assigned to us, but an assignment that constitutes our being. We exist *as* this responsibility; that is, to use Heidegger's term, we *ek-sist*, we are exposed to one another and together to the world that is nothing other than this exposition itself. Existence is responsibility for existence." Jean-Luc Nancy, "Responding to Existence," trans. Sarah Guyer, in *A Finite Thinking*, ed. Simon Sparks (Stanford University Press, 2003), 296.

26. On the dangerous corporealization of law in the modern era, see Ivan Nagel, *Autonomy and Mercy: Reflections on Mozart's Operas*, trans. Marion Faber and Ivan Nagel (Harvard University Press, 1991), 52–57, 145–48.

5. Hurt Feelings

1. Amanda Bailey and Mario DiGangi, "Introduction," in *Affect Theory and Early Modern Texts: Politics, Ecologies, and Form*, ed. Bailey and DiGangi (Palgrave Macmillan, 2017), 2; Steven Mullaney, *The Reformation of Emotions in the Age of Shakespeare* (University of Chicago Press, 2015), 6.

2. Gregory J. Seigworth and Melissa Gregg, "An Inventory of Shimmers," in *The Affect Theory Reader*, ed. Melissa Gregg and Gregory J. Seigworth (Duke University Press, 2010), 1.

3. Evelyn Tribble and John Sutton, "Cognitive Ecology as a Framework for Shakespearean Studies," in *Shakespeare and Emotion*, ed. Katherine A. Craik (Cambridge University Press, 2020), 101; Bailey and DiGangi, "Introduction," 4.

4. Embodiment's status as a sine qua non in early modern studies is suggested by Valerie Traub's editorial decision when confronted with the vast expansion of approaches addressing matters of sexuality—queer theory, race and ethnicity, empire, affect theory, disability, the inhuman—to rename a proposed collection on feminism and Shakespeare *The Oxford Handbook of Shakespeare and Embodiment.*

Her introduction effectively articulates the range of meanings that have attached themselves to the term. Valerie Traub, ed., *The Oxford Handbook of Shakespeare and Embodiment* (Oxford University Press, 2016), 32–35.

5. Gregg and Seigworth, "An Inventory of Shimmers," 9.

6. Brian Massumi, *Parables for the Virtual: Movement, Affect, Sensation* (Duke University Press, 2002), 28.

7. Bruce R. Smith, *Phenomenal Shakespeare* (Wiley-Blackwell, 2010), 25.

8. Smith, *Phenomenal Shakespeare,* 75, 80.

9. See Rei Terada, *Feeling in Theory: Emotion After the Death of the Subject* (Harvard University Press, 2003), 1–15.

10. Giulio J. Pertile, *Feeling Faint: Affect and Consciousness in the Renaissance* (Northwestern University Press, 2019), 151n22.

11. Pertile, *Feeling Faint,* 11.

12. Michel Henry, *Material Phenomenology* (Fordham University Press, 2008), 4. On the interruption and displacement inherent in feeling as "feeling oneself feel," see Jacques Derrida, *On Touching—Jean-Luc Nancy*, trans. Christine Irizarry (Stanford University Press, 2005), 31–38. For a subtle account of the incognizable nature of touch in Renaissance theater and the force involved in its literalization, see Carla Mazzio, "Acting with Tact: Touch and Theater in the Renaissance," in *Sensible Flesh: On Touch in Early Modern Culture*, ed. Elizabeth Harvey (University of Pennsylvania Press, 2002). See also Lynn Enterline's supple account of touch as indissociable from rhetorical suasion: "Afterword: Touching Rhetoric" in *Sensible Flesh*, ed. Harvey, 246–48.

13. Mark Hansen, "The Time of Affect: Bearing Witness to Life," *Critical Inquiry* 30, no. 3 (Spring 2004): 599.

14. Hansen, "The Time of Affect," 613.

15. Hansen, "The Time of Affect," 609.

16. Gail Kern Paster, "Communities: *Julius Caesar*," in *Shakespeare and Emotion*, ed. Craik, 99.

17. Paster, "Communities: *Julius Caesar*," 107, 99, 102.

18. Greg and Seigworth, "An Inventory of Shimmers," 1.

19. Paster, "Communities: *Julius Caesar*," 99.

20. Jean-Luc Nancy, *Being Singular Plural*, trans. Robert Richardson and Anne O'Byrne (Stanford University Press, 2000), 7.

21. Amanda Bailey, "Speak What We Feel: Sympathy and Statecraft," in *Affect Theory and Early Modern Texts*, ed. Bailey and DiGangi, 29–30, 36.

22. Terada, *Feeling in Theory*, 82.

23. Nancy, *Being Singular Plural*, 14.

24. Pertile, *Feeling Faint*, 135; Gregg and Seigworth, "An Inventory of Shimmers," 1, 2.

25. On embodiment as a motivated reification of linguistic polyvalency, see Marjorie Rubright, "Incorporating Kate: The Myth of Monolingualism in Shakespeare's *Henry the Fifth*," in *Oxford Handbook of Shakespeare and*

Embodiment, ed. Traub, 484–86; on embodiment as an effect of language's performative properties, see Stephen Spiess, "Puzzling Embodiment: Proclamation, *La Pucelle*, and *The First Part of Henry VI*," in *Oxford Handbook of Shakespeare and Embodiment*, ed. Traub, 93–95.

26. Mullaney, *The Reformation of Emotions*, 62.

27. Mullaney, *The Reformation of Emotions*, 7–8, 33.

28. Smith, *Phenomenal Shakespeare*, 156–57.

29. Evelyn Tribble and John Sutton, "Cognitive Ecology as a Framework for Shakespearean Studies," *Shakespeare Studies* 39 (2011): 97.

30. Tribble and Sutton, "Cognitive Ecology," 101.

31. Jane Bennett, *Vibrant Matter: A Political Ecology of Things* (Duke University Press, 2010), 121.

32. Pertile, *Feeling Faint*, 121. Understood in these terms, transcendence is never more assured than when we assume that relational networks will now let "the early moderns speak for themselves." Laurie Johnson, "*Cogito Ergo Theatrum*: Redistributing Cognition on the Early Modern Stage," in *Embodied Cognition and Shakespeare's Theater: The Early Modern Body-Mind*, ed. Laurie Johnson, John Sutton, and Evelyn Tribble (Routledge, 2014), 228.

33. As in, for example, Matthew Fuller and Olga Goriunova, *Bleak Joys: Aesthetics of Ecology and Impossibility* (University of Minnesota Press, 2019).

34. Nancy, *Being Singular Plural*, 33.

35. For a subtle, psychoanalytically oriented critique of totalizing accounts of environmentalism, see Steven Swarbrick, *The Environmental Unconscious: Ecological Poetics from Spenser to Milton* (University of Minnesota Press, 2023).

36. On theatricality as a historical interval between embodied spiritual emotion and embodied bourgeois affect, see Matthew Biberman, "Archeologies of Affect: Shakespeare's Phantasm of Love in *Romeo and Juliet*," in *Literature, Performance, and Somaesthetics: Studies in Agency and Embodiment*, ed. Katarzyna Lisowska and Jaroslaw Wozniak (Cambridge Scholars, 2017), 235–43.

37. A number of critics have offered affect-oriented readings of the play. On contentment as mediating term between psyche and community, see Paul Joseph Zajac, "The Politics of Contentment: Passions, Pastoral, and Community in Shakespeare's *As You Like It*," *Studies in Philology* 113, no. 2 (2016): 306–36; on the centrality of mutuality in the play, see Nathaniel Strout, "*As You Like It*, *Rosalynde*, and Mutuality," *Studies in English Literature* 41 (2001): 277–95; on the affective claims of rhetoric, see Swapan Chakravorty, "Translating Arden: Shakespeare's Rhetorical Place in *As You Like It*," in *Shakespeare and the Mediterranean*, ed. Susan Brock, Thomas Clayton, and Vincente Fores (University of Delaware Press, 2004), 156–67. See also Gail Kern Paster, *Humoring the Body: Emotions and the Shakespearean Stage* (University of Chicago Press, 2004), 118–25, who reads Rosalind's passage out of melancholy in terms of transformations in embodied humoreal temperature.

38. William Shakespeare, *As You Like It*, ed. Alan Brissenden (Oxford University

Press, 1993), 4.3.147–58. Hereafter cited parenthetically in the text by act, scene, and line numbers.

39. On the translation between those two modes in the play, see Albert Russell Ascoli, "Wrestling with Orlando: Chivalric Pastoral in Shakespeare's Arden," *Renaissance Drama* 36/37 (2010): 293–317.

40. On the play as a critique of the impulse to naturalize the fact of discursive mediation, see William N. West, *As If: Essays in 'As You Like It'* (Punctum, 2016), 25–31, 77–82; and Robert N. Watson, *Back to Nature: The Green and the Real in the Late Renaissance* (University of Pennsylvania Press, 2006), 78–107.

41. On the critical discomfort caused by the play's magical ending and the claim that it represents a recovery of the Golden Age promise of romance as against pastoral nostalgia, see Keir Elam, "'As They Did It in the Golden Age': Romantic Rapture and Semantic Rupture in *As You Like It*," *Canadian Review of Comparative Literature* 18, nos. 2–3 (1991): 229–31.

42. On a serpent's enfoldedness as a figure for reflexive consciousness, see Watson, *Back to Nature*, 85.

43. The scene in this sense embodies the creationist dimensions of the reflexive, liminal, and perpetually self-deviating character of the romance mode. On those traits, see Patricia Parker, *Inescapable Romance: Studies in the Poetics of a Mode* (Princeton University Press, 1979), 3–15.

44. Augustine, *St. Augustine's Confessions*, trans. Williams Watts (Heinemann, 1962), 20.

45. Sigmund Freud, *Mass Psychology and the Analysis of the "I,"* in *Mass Psychology and Other Writings*, trans. J. A. Underwood (Penguin, 2004), 75.

46. Freud, *Mass Psychology*, 78–79.

47. Freud, *Mass Psychology*, 78.

48. Freud, *Mass Psychology*, 59, 75.

49. "At its core," Lacan writes, desire is "desire for desire"; "man's desire is the desire of the Other." Jacques Lacan, *The Seminar of Jacques Lacan, Book X: Anxiety*, ed. Jacques-Alain Miller, trans. A. R. Price (Polity, 2014), 22, 332. The abyssal character of envy is implied by Freud's primary example of collective identification, a case of schoolgirls' hysterical identification with the anguished jealousy of a classmate: socialization originates, by such an account, as jealousy of jealousy.

50. Lacan, *The Seminar of Jacques Lacan, Book X*, 284.

51. Joannes Viterbiensis, *Liber de regimine civitatum*, ed. Caietano Salvimini (In Aedibus Successorum Monti, 1901), 276. Cited in Matthew G. Shoaf, "Eyeing Envy in the Arena Chapel," *Studies in Iconography* 30 (2009): 160n33. On envy as corrosive of communal bonds in the play, see Zajac, "Politics of Contentment," 326—corrosive of such bonds, and formative of them, we are suggesting.

52. On conversion conceived as going forward by turning backward to purer sources, see Jeffrey Shoulson, *Fictions of Conversion: Jews, Christians, and Cultures of Change in Early Modern England* (University of Pennsylvania Press, 2013), 83–85.

53. On fantasy as just such an event of constitutive interruption, see Jean

Laplanche and Jean-Bertrand Pontalis, "Fantasy and the Origins of Sexuality," *International Journal of Psycho-analysis* 49 (1968): 4–12.

54. In that sense, the scene anticipates the modern sense of "hurtling" as unreferenced or internally determined motion, as in a planet's movement through space. "Our president has always been out there. But on Jan. 6, 2021, he clearly reached escape velocity and hurtled into space." Jennifer Senior, "The Narcissist and Chief Brings It All Crashing Down," *New York Times*, January 10, 2021.

55. Troubling such boundaries, affect also troubles the distinctions that subtend accounts of affect as a life-substance "below" consciousness. Such oppositional accounts risk reinforcing the subjective structures they seek to undo.

56. A recursive logic equally played out at the play's end, as we noted in the last chapter. If we feel affronted when in the epilogue Rosalind, who has now let us know she's been only playing a woman, offers to kiss us, and if, caught thus, we feel compelled to mark a secure close to the drama with our applause, we will have had only ourselves to blame: She never said she was a man, only, "if I were a woman," the alternative to which is only ever, "if I were a man"—"much virtue in 'if,'" Touchstone declares toward the close. See Christopher Pye, "Introduction: Early Modern Political Aesthetics," in *Political Aesthetics in the Era of Shakespeare*, ed. Christopher Pye (Northwestern University Press, 2020), 13–16; and West, *As If*, 101–7.

57. On affect as redundant flight, see Marc Redfield, *The Politics of Aesthetics: Nationalism, Gender, Romanticism* (Stanford University Press, 2003), 113; and Terada, *Feeling in Theory*, 63.

58. See Heidegger's elaboration of "mit-Sein" in section 26 of *Being and Time*. Martin Heidegger, *Being and Time*, trans. Joan Stambaugh (State University of New York Press, 2010).

59. Nancy, *Being Singular Plural*, 8. For an especially deft account of *As You Like It*'s commitment to the promise of other worlds, see West, *As If*, 77–82, 107–9.

60. Walter Benjamin, "Theses on the Philosophy of History," in *Illuminations: Essays and Reflections*, ed. Hannah Arendt (Harcourt, 1968), 257.

61. Benjamin, "Theses on the Philosophy of History," 258.

62. Benjamin, "Theses on the Philosophy of History," 257.

6. The Spider and the Spinning Wheel

1. Cited in Gary Tinterow, "Raphael Replaced: The Triumph of Spanish Painting in France," in *Manet/Velázquez: The French Taste for Spanish Painting*, ed. Gary Tinterow and Genevieve Lacambre (Yale University Press, 2003), 3.

2. Kenneth Clark, "Six Great Pictures: 'Las Meninas,'" *The Sunday Times* (London), June 2, 1957.

3. Jean-Luc Nancy, *The Ground of the Image*, trans. Jeff Fort (Fordham University Press, 2005), 80. On the proto-Kantian dimensions of Velázquez's art—its dethroning of the object as given—see Daniel Arasse, *Take a Closer Look* (Princeton University Press, 2013), 119–24.

4. Giles Knox, *The Late Paintings of Velázquez: Theorizing Painterly Performance* (Ashgate, 2009), 72.

5. Knox, *The Late Paintings of Velázquez*, 73.

6. On the painting as a rivalrous homage to Titian and Rubens, see Jonathan Brown and Carmen Garrido, *Velázquez: The Techniques of Genius* (Yale University Press, 1998), 200; Giles Knox, *Sense Knowledge and the Challenge of Italian Renaissance Art* (Amsterdam University Press, 2019), 89.

7. The female spinners repeat the postures of Michelangelo's Sistine Chapel *ignudi*. Knox, *The Late Paintings of Velázquez*, 99. On contemporary debates concerning the opposition between conceptual and tactile conceptions of artistic practice, see Knox, *The Late Paintings of Velázquez*, 84–87; Knox, *Sense Knowledge*, 82–90; Svetlana Alpers, "Interpretation Without Representation, or, the Viewing of *Las Meninas*," *Representations* 1 (February 1983): 33–34; Richard Pym, "Interdiction of Closure in Velázquez's 'Fable,'" *Iberian and Latin American Studies* 5, no. 2 (1999): 189–90.

8. Pliny, *Natural History*, cited in Jonathan Brown and Carmen Garrido, *Velázquez: The Techniques of Genius*, 199. The text was in Velázquez's library. Antiphilus, meanwhile, was praised for his representation of a "spinning room, in which with great speed the women are working at their duties." Cited in Knox, *The Late Paintings of Velázquez*, 71.

9. Svetlana Alpers, *The Vexations of Art: Velázquez and Others* (Yale University Press, 2005), 153.

10. Alpers, *The Vexations of Art*, 151. On the shifting fate of the mechanical arts over the course of the late Middle Ages and Renaissance, see David Summers, *The Judgment of Sense: Renaissance Naturalism and the Rise of Aesthetics* (Cambridge University Press, 1990), 316.

11. Jan Baptiste Bedaux, "Velázquez's Fable of Arachne: A Continuing Story," *Simiolus* 21, no. 4 (1992): 304–5.

12. On the former, see "Titian's Rape of Europa," Gardner Museum online commentary; on the latter, see Nancy, *The Ground of the Image*, 128.

13. Mary Crawford Volk, "On Velázquez and the Liberal Arts," *Art Bulletin* 60 (1978): 286.

14. Ovid, *Metamorphoses*, trans. Rolfe Humphries (Indiana University Press, 2018), 55.

15. Ovid, *Metamorphoses*, 130.

16. According to Velázquez's contemporary Marco Boschini, the *colpo*, or thrust of the brush, amounted to "intelligence itself." Marco Boschini, *La carta del navegar pitoresco*, ed. Pallucchini (Venice, 1660). Cited in Knox, *The Late Paintings of Velázquez*, 86.

17. "The sensible owes its being to its division," Nancy writes. Jean-Luc Nancy, *The Sense of the World*, trans. Jeffrey Librett (University of Minnesota Press, 2008), 129.

18. "The image contains the index of its frozenness (its form, its present, its

representation) and at the same time the index of movement (force, appearing/disappearing)." Nancy, *The Ground of the Image*, 98.

19. Knox, *Sense Knowledge*, 103; Jonathan Brown, *Velázquez: Painter and Courtier* (Yale University Press, 1986), 75.

20. "Velázquez achieved two seemingly contradictory aims—one is a heightened realism, the other a richness and variety of surface that insistently calls itself to our attention." Brown, *Velázquez*, 148.

21. Alpers, *The Vexations of Art*, 154.

22. Maurice Blanchot, "Athenaeum," *Studies in Romanticism* 27, no. 2 (1983): 165.

23. On that reflexive mechanism, see Marc Redfield, *The Politics of Aesthetics: Nationalism, Gender, Romanticism* (Stanford University Press, 2003), 10; and Christopher Pye, *The Storm at Sea: Political Aesthetics in the Time of Shakespeare* (Fordham University Press, 2015), 14–16.

24. Aristotle, *Politics*, trans. Harris Rackham, Loeb Classics (Harvard University Press, 1944), book 1, para. 7. I owe the citation, and my sense of its pertinence in this context, to Ted Tregear.

25. On the animal as a necessary and unmasterable figure of finitude and difference, see Jacques Derrida, *The Animal That Therefore I Am* (Fordham University Press, 2009).

26. On temporal delay as the core element in autopoietic formation, see Niklas Luhmann, *Introduction to Systems Theory*, trans. Peter Gilgen (Polity, 2013), 52, 142–43.

27. Veronique Foti, "Representation Represented: Foucault, Velázquez, Descartes," *Postmodern Culture* 7, no. 1 (September 1996): 32–33. For Derrida, that threat of uncategorizably multiple otherness lies at the heart of the defensive impulse to reduce the nonhuman life forms to a single term, "the animal."

28. On touch as the possibility of being touched, an "acceding without access," see Nancy, *The Sense of the World*, 83, 128.

29. Marc Redfield, *Phantom Formations: Aesthetic Ideology and the Bildungsroman* (Cornell University Press, 1996), 197.

30. Foti provides a subtle account of the way in which Foucault's geometrical orientation "does not give due weight to the materiality" of Velázquez's practice and, thus, to the degree to which "one's perceptions may be part and parcel of the procession of illusions and simulacra." Foti, "Representation Represented," 31, 32–33.

31. Joel Snyder and Ted Cohen, "Reflexions on 'Las Meninas': Paradox Lost," *Critical Inquiry* 7, no. 2 (1980): 446–47.

32. Arasse, *Take a Closer Look*, 110.

33. Brown, *Velázquez*, 260.

34. Hubert Damisch, *The Origin of Perspective*, trans. John Goodman (MIT Press, 1995), 438.

35. Jacques Lacan, *The Four Fundamental Concepts of Psychoanalysis*, trans. Alan Sheridan (Norton, 1998), 88–89.

36. Damisch, *The Origin of Perspective*, 438.

37. Jacques Lacan, *Seminar 13: L'objet*, Lesson 22, June 15, 1966, 270–74, http://staferla.free.fr/. See Tom Brockelman, "The Other Side of the Canvas: Lacan Flips Foucault," *Continental Philosophy Review* 46, no. 2 (2013): 280.

38. Jacques Lacan, *Seminar 13: L'objet*, 276.

39. On the conflicting accounts, see Arasse, "Take a Closer Look," 110.

40. Brown, *Velázquez*, 259–60.

41. On the temporal lapse and lost origination implied by the turned canvas, see Brockelman, "The Other Side of the Canvas," 283.

42. On the motif of the suspended hand as implying art as an intellectual rather than manual act, see John Moffitt, "Velázquez in the Alcazar Palace, 1656," *Art History* 6, no. 3 (1983): 291–94.

43. On that merging, see Knox, *Sense Knowledge*, 96.

44. On the parallel between the "recalcitrant canvas" and the dog "peacefully relinquishing itself to just being seen," see Foti, "Representation Represented," 15.

45. Michel Foucault, *Les mots et les choses* (Gallimard, 1966), 25.

46. Jacques Lacan, *Seminar 13: L'objet*, 234.

47. In one way or another the aberrant readings forwarded by Foucault and Lacan can be understood in terms of Paul de Man's analysis of the proper name as a form of forced anthropomorphic identification necessary to and disruptive of meaning as a coherent system of tropes. Apostrophe lays bare the arbitrariness, and thus the force, involved in that anthropomorphic imperative. See Paul de Man, "Anthropomorphism and Trope in the Lyric," in *The Rhetoric of Romanticism* (Columbia University Press, 1984), 239–43; and Paul de Man, "Hypogram and Inscription," in *The Resistance to Theory* (University of Minnesota Press, 1986), 47–50.

48. On the parergonal frame as phenomenal/nonphenomenal "cut," see Jacques Derrida, *The Truth in Painting*, trans. Geoffrey Bennington (University of Chicago Press, 1987), 89–103; and Redfield, *Phantom Formations*, 17.

49. Foti, "Representation Represented," 14.

50. Dario Gamboni, *The Destruction of Art: Iconoclasm and Vandalism Since the French Revolution* (Reaktion, 1997), 94.

Bibliography

Adelman, Janet. *Blood Relations: Christian and Jew in 'The Merchant of Venice.'* University of Chicago Press, 2008.

———. *The Common Liar: An Essay on 'Antony and Cleopatra.'* Yale University Press, 1973.

Agamben, Giorgio. *Homo Sacer: Sovereign Power and Bare Life*. Trans. Daniel Heller-Roazen. Stanford University Press, 1998.

———. *The Kingdom and the Glory*. Trans. Lorenzo Chiesa. Stanford University Press, 2011.

———. *The Man Without Content*. Trans. Georgia Albert. Stanford University Press, 1999.

———. *The Time That Remains: A Commentary on the Letter to the Romans*. Trans. Patricia Dailey. Stanford University Press, 2005.

———. *The Use of Bodies*. Trans. Adam Kotsko. Stanford University Press, 2016.

Alpers, Svetlana. "Interpretation Without Representation, or, The Viewing of *Las Meninas*." *Representations* 1 (February 1983): 31–42.

———. *The Vexations of Art: Velázquez and Others*. Yale University Press, 2005.

Arasse, Daniel. *Take a Closer Look*. Princeton University Press, 2013.

Aristotle. *The Poetics*. Trans. S. H. Butcher. Cosimo, 2008.

———. *Politics*. Trans. Harris Rackham. Loeb Classics. Harvard University Press, 1944.

Ascoli, Albert Russell. "Wrestling with Orlando: Chivalric Pastoral in Shakespeare's Arden." *Renaissance Drama* 36/37 (2010): 293–317.

Augustine. *Saint Augustine's Confessions*. Trans. Williams Watts. Heinemann, 1962.

Badiou, Alain. *Being and Event*. Trans. Oliver Feltham. Continuum, 2005.

Bailey, Amanda. "Speak What We Feel: Sympathy and Statecraft." In *Affect Theory and Early Modern Texts: Politics, Ecologies, and Form*, ed. Amanda Bailey and Mario DiGangi, 27–46. Palgrave Macmillan, 2017.

Bailey, Amanda, and Mario DiGangi. "Introduction." In *Affect Theory and Early Modern Texts: Politics, Ecologies, and Form*, ed. Amanda Bailey and Mario DiGangi. Palgrave Macmillan, 2017.

Barber, C. L. *Shakespeare's Festive Comedies*. Princeton University Press, 2012.

Barret, J. K. "The Crowd in Imogen's Bedroom: Allusion and Ethics in *Cymbeline*." *Shakespeare Quarterly* 66, no. 4 (2015): 440–62.

Bedaux, Jan Baptiste. "Velázquez's Fable of Arachne: A Continuing Story." *Simiolus* 21, no. 4 (1992): 296–305.

Belsey, Catherine. "Cleopatra's Seduction." In *Alternative Shakespeares 2*, ed. Terence Hawkes. Routledge, 1996.

Benjamin, Walter. "One Way Street." In *Selected Writings*, vol. 1: *1913–1926*, ed. Marcus Bullock and Michael W. Jennings, 444–88. Harvard University Press, 1996.

———. "Theses on the Philosophy of History." In *Illuminations: Essays and Reflections*, ed. Hannah Arendt. Harcourt, 1968.

Bennett, Jane. *Vibrant Matter: A Political Ecology of Things*. Duke University Press, 2010.

Berger, Harry. "Mercifixion in *The Merchant of Venice*." *Renaissance Drama* 38, no. 1 (2010): 3–45.

Biberman, Matthew. "Archeologies of Affect: Shakespeare's Phantasm of Love in *Romeo and Juliet*." In *Literature, Performance, and Somaesthetics: Studies in Agency and Embodiment*, ed. Katarzyna Lisowska and Jaroslaw Wozniak. Cambridge Scholars, 2017.

Billings, Joshua. *Genealogy of the Tragic: Greek Tragedy and German Philosophy*. Princeton University Press, 2014.

Blanchot, Maurice. "Athenaeum." *Studies in Romanticism* 27, no. 2 (1983): 163–72.

Boschini, Marco. *Carta del navegar pitoresco*. Ed. Anna Pallucchini. Torino: Loescher, 1660.

Bogost, Ian. *Alien Phenomenology, or What It's Like to Be a Thing*. University of Minnesota Press, 2012.

Braidotti, Rosie. "The Politics of 'Life Itself' and New Ways of Dying." In *New Materialisms: Ontology, Agency, and Politics*, ed. Diana Coole and Samantha Frost, 201–18. Duke University Press, 2010.

Brill, Lesley. "Truth and *Timon of Athens*." *Modern Language Quarterly* 40, no. 1 (1979): 17–36.

Brockelman, Tom. "The Other Side of the Canvas: Lacan Flips Foucault." *Continental Philosophy Review* 46, no. 2 (2013): 271–90.

Brown, Jonathan. *Velázquez: Painter and Courtier*. Yale University Press, 1986.

Brown, Jonathan, and Carmen Garrido. *Velázquez: The Techniques of Genius*. Yale University Press, 1998.

Campana, Joseph. "On Not Defending Poetry: Spenser, Suffering, and the Energy of Affect." *PMLA* 120, no. 1 (2005): 33–48.

Campana, Joseph, and Scott Maisano. "Introduction." In *Renaissance Posthumanism*, ed. Joseph Campana and Scott Maisano, 1–36. Fordham University Press, 2016.

Chakravorty, Swapan. "Translating Arden: Shakespeare's Rhetorical Place in *As You Like It*." In *Shakespeare and the Mediterranean*, ed. Susan Brock, Thomas Clayton, and Vincente Fores, 156–67. University of Delaware Press, 2004.

Cheah, Pheng. *What Is a World? On Postcolonial Literature as World Literature*. Duke University Press, 2016.

Clark, Kenneth. "Six Great Pictures: 'Las Meninas.'" *Sunday Times* (London), June 2, 1957.

Colebrook, Claire. *Death of the Posthuman: Essays on Extinction*. Vol. 1. Open Humanities Press, 2015.

Damisch, Hubert. *The Origin of Perspective*. Trans. John Goodman. MIT Press, 1995.

Daniel, Drew. *Joy of the Worm: Suicide and Pleasure in Early Modern English Literature*. University of Chicago Press, 2022.

Darcy, Robert. "Shakespeare's Empty Plot: The Epicenotaph in *Timon of Athens*." *Renaissance Drama* 33 (2004): 159–79.

Davis, Natalie Z. *The Return of Martin Guerre*. Harvard University Press, 1983.

Dawson, Brent. "Multiple Worlds in *Antony and Cleopatra*." *Renaissance Drama* 43 (2015): 173–91.

———. "Worldly Muck: Translating Matter in Book 2 of *The Faerie Queene*." In *This Distracted Globe*, ed. Jonathan Goldberg, Karen Newman, and Marcie Frank, 23–44. Fordham University Press, 2016.

Degenhardt, Jane Hwang. "Counter-Imperial Horizons in *Antony and Cleopatra*." *Studies in English Literature 1500–1900* 62, no. 1 (2022): 47–76.

Degenhardt, Jane Hwang, and Henry Turner. "Thinking with Cosmology: Aesthetics, Worlds, and Kinds in Shakespeare's *The Tempest*." In *Shakespearean Cosmologies*. Forthcoming.

de Man, Paul. *Aesthetic Ideology*. Ed. Andrzej Warminski. University of Minnesota Press, 1966.

———. *The Rhetoric of Romanticism*. Columbia University Press, 1984.

Derrida, Jacques. *The Animal That Therefore I Am*. Trans. David Wills. Fordham University Press, 2008.

———. *The Beast and the Sovereign*. Vol. 2. University of Chicago Press, 2011.

———. "Before the Law." In *Acts of Literature*, ed. Derek Attridge. Routledge, 1992.

———. "The Ends of Man." In *Margins of Philosophy*, trans. Alan Bass, 111–36. University of Chicago Press, 1972.

———. "Force of Law: 'The Mystical Foundation of Authority.'" In *Deconstruction and the Possibility of Justice*, ed. Drucilla Cornell, Michel Rosenfeld, and David Gray Carlson, 3–67. Routledge, 1992.

———. *On Touching—Jean-Luc Nancy*. Trans. Christine Irizarry. Stanford University Press, 2005.

———. "Signature, Event, Context." In *Margins of Philosophy*, trans. Alan Bass, 307–21. University of Chicago Press, 1982.

———. *Specters of Marx: The State of the Debt, the Work of Mourning, and the New International*. Trans. Peggy Kamuf. Routledge, 1994.

———. *The Truth in Painting*. Trans. Geoffrey Bennington. University of Chicago Press, 1987.

Dionne, Craig. *Posthuman Lear*. Punctum, 2016.

Eden, Kathy. *Poetic and Legal Fiction in the Aristotelian Tradition*. Princeton University Press, 1986.

Elam, Keir. "'As They Did It in the Golden Age': Romantic Rapture and Semantic Rupture in *As You Like It*." *Canadian Review of Comparative Literature* 18, nos. 2–3 (1991).

Elden, Stuart. *Shakespearean Territories*. University of Chicago Press, 2018.

Enterline, Lynn."Afterword: Touching Rhetoric." In *Sensible Flesh: On Touch in Early Modern Culture*, ed. Elizabeth Harvey, 243–54. University of Pennsylvania Press, 2002.

Esposito, Roberto. *Communitas: The Origin and Destiny of Community*. Trans. Timothy Cambell. Stanford University Press, 2010.

Evans, Meredith. "Cities of the Stranger." In *This Distracted Globe: Worldmaking in Early Modern Europe*, ed. Jonathan Goldberg, Karen Newman, and Marcie Frank, 142–60. Fordham University Press, 2016.

Fichter, Andrew. "*Antony and Cleopatra*: 'The Time of Universal Peace.'" *Shakespeare Survey* 33 (1980): 99–112.

Floyd-Wilson, Mary, and Garrett A. Sullivan Jr. "Introduction: Inhabiting the Body, Inhabiting the World." In *Environment and Embodiment in Early Modern Drama and Performance*, ed. Mary Floyd-Wilson and Garrett A. Sullivan Jr., 1–13. Palgrave, 2007.

Foti, Veronique. "Representation Represented: Foucault, Velázquez, Descartes." *Postmodern Culture* 7, no. 1 (September 1996).

Foucault, Michel. *Les mots et les choses*. Gallimard, 1966.

Freud, Sigmund. *Mass Psychology and the Analysis of the "I."* In *Mass Psychology and Other Writings*, trans. J. A. Underwood. Penguin, 2004.

Frye, Northrop. *The Anatomy of Criticism: Four Essays*. Princeton University Press, 2020.

———. *A Natural Perspective: The Development of Shakespearean Comedy and Romance*. Columbia University Press, 1965.

Fuller, Matthew, and Olga Goriunova. *Bleak Joys: Aesthetics of Ecology and Impossibility*. University of Minnesota Press, 2019.

Gamboni, Dario. *The Destruction of Art: Iconoclasm and Vandalism Since the French Revolution*. Reaktion, 1997.

Gearhart, Suzanne. *The Interrupted Dialectic: Philosophy, Psychoanalysis, and Their Tragic Others*. Johns Hopkins University Press, 1992.

Gillies, John. *Shakespeare and the Geography of Difference*. Cambridge University Press, 1994.

Glimp, David. "Extreme Cary." In *This Distracted Globe*, ed. Jonathan Goldberg, Karen Newman, and Marie Frank, 45–63. Fordham University Press, 2016.

Goldberg, Jonathan, Karen Newman, and Marcie Frank. "Introduction: World Enough and Time." In *This Distracted Globe: Worldmaking in Early Modern Europe*, edited by Jonathan Goldberg, Karen Newman, and Marcie Frank, 1–18. Fordham University Press, 2016.

Greenblatt, Stephen. "Psychoanalysis and Renaissance Culture." In *Learning to Curse: Essays in Early Modern Culture*. Routledge, 1990.

Greene, Thomas M. *The Light in Troy: Imitation and Discovery in Renaissance Poetry*. Yale University Press, 1986.

Hamacher, Werner. "'Now': Walter Benjamin on Historical Time." In *The Moment: Time and Rupture in Modern Thought*, ed. Heidrun Friese. Liverpool University Press, 2001.

Hansen, Mark. "The Time of Affect: Bearing Witness to Life." *Critical Inquiry* 30, no. 3 (Spring 2004): 584–626.

Harris, Jonathan Gil. "'Narcissus in Thy Face': Roman Desire and the Difference It Fakes in *Antony and Cleopatra*." *Shakespeare Quarterly* 45, no. 4 (1994): 408–25.

Hartog, François. *Regimes of Historicity: Presentism and Experiences of Time*. Trans. Saskia Brown. Columbia University Press, 2003.

Hawkes, Terence. *Shakespeare in the Present*. Routledge, 2002.

Hegel, G. W. F. *The Phenomenology of Spirit*. Trans. Terry Pinkard. Cambridge University Press, 2018.

Heidegger, Martin. *Being and Time*. Trans. Joan Stambaugh. State University of New York Press, 2010.

———. *Introduction to Metaphysics*. Trans. Gregory Fried and Richard Polt. Yale University Press, 2000.

———. "The Question Concerning Technology." In *Basic Writings*, trans. David Krell. New York: HarperCollins, 2008.

———. "The Thing." In *Poetry, Language, Thought*, trans. Albert Hofstadter. HarperCollins, 1971.

Henry, Michel. *Material Phenomenology*. Trans. Scott Davidson. Fordham University Press, 2008.

Hutson, Lorna. *The Invention of Suspicion: Law and Mimesis in Shakespeare and Renaissance Drama*. Oxford University Press, 2007.

Hyman, Wendy Beth. "'Beyond Beyond': *Cymbeline*, the Camera Obscura, and the Ontology of Elsewhere." *English Literary Renaissance* 52, no. 3 (2022): 397–412.

Irigaray, Luce. *This Sex Which Is Not One*. Trans. Catherine Porter. Cornell University Press, 1983.

Jackson, Ken. "One Wish or the Possibility of the Impossible: Derrida, the Gift, and God in *Timon of Athens*." *Shakespeare Quarterly* 52, no. 1 (2001): 34–66.

Jameson, Frederick. *The Political Unconscious: Narrative as a Socially Symbolic Act*. Routledge, 1981.

Johnson, Laurie. "*Cogito Ergo Theatrum*: Redistributing Cognition on the Early Modern Stage." In *Embodied Cognition and Shakespeare's Theater: The Early Modern Body-Mind*, ed. Laurie Johnson, John Sutton, and Evelyn Tribble, 216–33. Routledge, 2014.

Johnson, Lynn. "Friendship, Coercion, and Interest: Debating the Foundation of Justice in Early Modern England." *Journal of Early Modern History* 8 (2004): 46–64.

Junker, William. "The Image of Both Theaters: Empire and Revelation in Shakespeare's *Antony and Cleopatra*." *Shakespeare Quarterly* 66, no. 2 (Summer 2015): 167–87.

Kahn, Coppelia. "'Magic of Bounty': *Timon of Athens*, Jacobean Patronage, and Maternal Power." *Shakespeare Quarterly* 38, no. 1 (1987): 34–57.

Kant, Immanuel. *Der Streit der Fakultaten*. In *Kant's Werke: Akademie Textausgabe*. Gruyter, 1968.

Knox, Giles. *The Late Paintings of Velázquez: Theorizing Painterly Performance*. Ashgate, 2009.

———. *Sense Knowledge and the Challenge of Italian Renaissance Art*. Amsterdam University Press, 2019.

Kolb, Laura. "Debt's Poetry in *Timon of Athens*." *Studies in English Literature* 58, no. 2 (2018): 399–419.

Koselleck, Reinhart. *Futures Past: On the Semantics of Historical Time*. Trans. Keith Tribe. Columbia University Press, 2004.

Kottman, Paul A. "Defying the Stars: Tragic Love as the Struggle for Freedom in *Romeo and Juliet*." *Shakespeare Quarterly* 63, no. 1 (2012): 1–38.

Lacan, Jacques. *The Four Fundamental Concepts of Psychoanalysis*. Ed. Jacques-Alain Miller. Trans. Alan Sheridan. Norton, 1977.

———. *The Seminar of Jacques Lacan, Book X: Anxiety*. Ed. Jacques-Alain Miller. Trans. A. R. Price. Polity Press, 2014.

———. *The Seminar of Jacques Lacan. Book XIII: L'objet*. http://staferla.free.fr.

Laclau, Ernesto. *The Rhetorical Foundations of Society*. Verso, 2014.

Lacoue-Labarthe, Philippe. *Poetics of History: Rousseau and the Theater of Originary Mimesis*. Trans. Jeff Fort. Fordham University Press, 2019.

———. *Typography: Mimesis, Philosophy, Politics*. Trans. Christopher Fynsk. Harvard University Press, 1989.

Lamb, Mary Ellen, and Valerie Wayne. "Introduction: Into the Forest." In *Staging Early Modern Romance: Prose Fiction, Dramatic Romance, and Shakespeare*, ed. Mary Ellen Lamb and Valerie Wayne. Routledge, 2009.

Lammer, Lukas. *Shakespearean Temporalities: History on the Early Modern Stage*. Routledge, 2018.

Laplanche, Jean, and Jean-Bertrand Pontalis. "Fantasy and the Origins of Sexuality." *International Journal of Psycho-analysis* 49 (1968): 1–18.

Latour, Bruno. *Reassembling the Social: An Introduction to Actor-Network Theory*. Oxford University Press, 2007.

Lefort, Claude. *Democracy and Political Theory*. Trans. David Macey. University of Minnesota Press, 1988.

Leo, Russ. *Tragedy as Philosophy in the Reformation World*. Oxford University Press, 2019.

Linton, Joan Pong. "Timon's Hunger in the Forest: Toward a Political Aesthetics of Being Beside Oneself." In *Political Aesthetics in the Era of Shakespeare*, ed. Christopher Pye. Northwestern University Press, 2020.

Loomba, Ania. *Shakespeare, Race, and Colonialism*. Oxford University Press, 2002.

Luhmann, Niklas. *Art as Social System*. Trans. Eva M. Knodt. Stanford University Press, 2000.

———. *Introduction to Systems Theory*. Trans. Peter Gilgen. Polity, 2013.

———. *Observations on Modernity*. Trans. William Whobrey. Stanford University Press, 1998.

Lupton, Julia. *Afterlives of the Saints: Hagiography, Typology, and Renaissance Literature*. Stanford University Press, 1996.

———. *Thinking with Shakespeare*. University of Chicago Press, 2011.

MacDonald, Ronald R. "Playing Til Doomsday: Interpreting *Antony and Cleopatra*." *Studies in the English Renaissance* 15, no. 1 (1985): 78–99.

Maley, Willy. "Cymbeline, the Font of History, and the Matter of Britain." In *Alternative Shakespeares* 3, ed. Diana Henderson, 119–37. Routledge, 2008.

Mapp, Nigel. "Early Modern Aesthetics: *Antony and Cleopatra* and the Afterlife of Domination." *Aisthesis* 13, no. 2 (2020): 169–84.

Marx, Karl. *Capital*. Trans. Eden Paul and Cedar Paul. Dent, 1939.

Massumi, Brian. *Parables for the Virtual: Movement, Affect, Sensation*. Duke University Press, 2002.

Mazzio, Carla. "Acting with Tact: Touch and Theater in the Renaissance." In *Sensible Flesh: On Touch in Early Modern Culture*, ed. Elizabeth Harvey, 159–86. University of Pennsylvania Press, 2002.

McEachern, Claire. *The Politics of English Nationhood: 1590–1612*. Cambridge University Press, 1996.

Mikalachki, Jodi. "The Masculine Romance of Roman Britain: Cymbeline and Early Modern English Nationalism." In *Shakespeare's Romances*, ed. Alison Thorne, 301–22. Palgrave, 2003.

Moffitt, John. "Velázquez in the Alcazar Palace, 1656." *Art History* 6, no. 3 (1983): 271–300.

Montaigne, Michel de. *The Complete Essays of Montaigne*. Trans. Donald Frame. Stanford University Press, 1976.

Mullaney, Steven. *The Reformation of Emotions in the Age of Shakespeare*. University of Chicago Press, 2015.

Nagel, Ivan. *Autonomy and Mercy: Reflections on Mozart's Operas*. Trans. Marion Faber and Ivan Nagel. Harvard University Press, 1991.

Nancy, Jean-Luc. *After Fukushima: The Equivalence of Disasters*. Trans. Charlotte Mandell. Fordham University Press, 2014.

———. *Being Singular Plural*. Trans. Robert Richardson and Anne O'Byrne. Stanford University Press, 2000.

———. *The Creation of the World or Globalization*. Trans. François Raffoul and David Pettigrew. State University of New York Press, 2007.

———. *A Finite Thinking*. Ed. Simon Sparks. Stanford University Press, 2003.

———. *The Ground of the Image*. Trans. Jeff Fort. Fordham University Press, 2005.

———. *The Sense of the World*. Trans. Jeffrey Librett. University of Minnesota Press, 2008.

Ovid. *Metamorphoses*. Trans. Rolfe Humphries. Indiana University Press, 2018.

Palfrey, Simon. *Late Shakespeare: A New World of Words*. Oxford University Press, 1997.

———. *Shakespeare's Possible Worlds*. Cambridge University Press, 2017.

Parker, Patricia. *Inescapable Romance: Studies in the Poetics of a Mode*. Princeton University Press, 1979.

Paster, Gail Kern. "Communities: *Julius Caesar*." In *Shakespeare and Emotion*, ed. Katherine A. Craik, 94–108. Cambridge University Press, 2020.

———. *Humoring the Body: Emotions and the Shakespearean Stage*. University of Chicago Press, 2004.

Patterson, Annabel. *Shakespeare and the Popular Voice*. Blackwell, 1989.

Pertile, Giulio J. *Feeling Faint: Affect and Consciousness in the Renaissance*. Northwestern University Press, 2019.

Pye, Christopher. "Green World / No World: *King Lear*, Eco-Criticism, and the Politics of Finitude." *Shakespeare Jahrbuch* 153 (2017): 11–34.

———. "Introduction." In *Political Aesthetics in the Era of Shakespeare*, ed. Christopher Pye, 1–23. Northwestern University Press, 2020.

———. "'No Cause, No Cause': *King Lear* and the Space of the World." In *Political Aesthetics in the Era of Shakespeare*, ed. Christopher Pye, 45–66. Northwestern University Press, 2020.

———. *The Storm at Sea: Political Aesthetics in the Time of Shakespeare*. Fordham University Press, 2015.

Pym, Richard. "Interdiction of Closure in Velázquez's 'Fable.'" *Iberian and Latin American Studies* 5, no. 2 (1999): 189–99.

Quint, David. *Epic and Empire: Politics and Generic Form from Virgil to Milton*. Princeton University Press, 1993.

Raber, Karen. *Shakespeare and Posthumanist Theory*. Arden, 2018.
Ramachandran, Ayesha. "How to Theorize the World: An Early Modern Manifesto." *New Literary History* 48, no. 4 (2017): 655–84.
———. *The Worldmakers: Global Imagining in Early Modern Europe*. University of Chicago Press, 2018.
Rancière, Jacques. *Aesthetics and Its Discontents*. Polity, 2009.
Redfield, Marc. *Phantom Formations: Aesthetic Ideology and the Bildungsroman*. Cornell University Press, 1996.
———. *The Politics of Aesthetics: Nationalism, Gender, Romanticism*. Stanford University Press, 2003.
Robertson, Lauren. *Entertaining Uncertainty in the Early Modern Theater: Stage Spectacle and Audience Response*. Cambridge University Press, 2023.
Ronen, Ruth. *Possible Worlds in Literary Theory*. Cambridge University Press, 2009.
Rubright, Marjorie. "Incorporating Kate: The Myth of Monolingualism in Shakespeare's *Henry the Fifth*." In *Oxford Handbook of Shakespeare and Embodiment*, ed. Valerie Traub, 468–90. Oxford University Press, 2016.
Sands, Kathleen M. "Tragedy, Theology, and Feminism in the Time After Time." In *Rethinking Tragedy*, ed. Rita Felski, 41–61. Johns Hopkins University Press, 2008.
Schmidt, Dennis J. *On Germans and Other Greeks: Tragedy and Ethical Life*. Indiana University Press, 2001.
Schmitt, Carl. *The Nomos of the Earth in the International Law of Jus Publicum Europaeum*. Trans. G. L. Ulmen. Telos, 2006.
Scholar, Richard. *The* Je-Ne-Sais-Quoi *in Early Modern Europe*. Oxford University Press, 2005.
Seigworth, Gregory J., and Melissa Gregg. "An Inventory of Shimmers." In *The Affect Theory Reader*, ed. Melissa Gregg and Gregory J. Seigworth, 1–26. Duke University Press, 2010.
Senior, Jennifer. "The Narcissist and Chief Brings It All Crashing Down." *New York Times*, January 10, 2021.
Shakespeare, William. *Anthony and Cleopatra*. Ed. Michael Neil. Oxford University Press, 1994.
———. *As You Like It*. Ed. Alan Brissenden. Oxford University Press, 1993.
———. *Cymbeline*. In *The Norton Shakespeare*, 3rd ed., ed. Stephen Greenblatt et al. Norton, 2016.
———. *King Lear*. Ed. R. A. Foakes. Arden Shakespeare, 1997.
———. *The Merchant of Venice*. In *The Norton Shakespeare*, 3rd ed., ed. Stephen Greenblatt et al. Norton, 2016.
———. *A Midsummer Night's Dream*. In *The Norton Shakespeare*, 3rd ed., ed. Stephen Greenblatt et al. Norton, 2016.
———. *Othello*. Ed. Michael Neill. Oxford University Press, 2008.

———. *Pericles*. In *The Norton Shakespeare*, 3rd ed., ed. Stephen Greenblatt et al. Norton, 2016.
———. *The Tempest*. Ed. Stephen Orgel. Clarendon, 1987.
———. *Twelfth Night*. In *The Norton Shakespeare*, 3rd ed., ed. Stephen Greenblatt et al. Norton, 2016.
———. *The Winter's Tale*. In *The Norton Shakespeare*, 3rd ed., ed. Stephen Greenblatt et al. Norton, 2016.
Shakespeare, William, and Thomas Middleton. *Timon of Athens*. Ed. John Jowett. Oxford University Press, 2004.
Shepherdson, Charles. "Aesthetic 'Sense' in Kant and Nancy." *New Literary History* 48 (2017): 197–221.
Shoaf, Matthew G. "Eyeing Envy in the Arena Chapel." *Studies in Iconography* 30 (2009): 126–67.
Shohet, Lauren. *Temporality, Genre, and Experience in the Age of Shakespeare: Forms of Time*. Bloomsbury, 2018.
Shoulson, Jeffrey. *Fictions of Conversion: Jews, Christians, and Cultures of Change in Early Modern England*. University of Pennsylvania Press, 2013.
Smith, Bruce R. *Phenomenal Shakespeare*. Wiley-Blackwell, 2010.
Snyder, Joel, and Ted Cohen. "Reflexions on 'Las Meninas': Paradox Lost." *Critical Inquiry* 7, no. 2 (1980): 429–47.
Spiess, Stephen. "Puzzling Embodiment: Proclamation, La Pucelle, and the First Part of Henry VI." In *The Oxford Handbook of Shakespeare and Embodiment*, ed. Valerie Traub, 93–111. Oxford University Press, 2016.
Spinosa, Charles. "The Transformation of Intentionality: Debt and Contract in *The Merchant of Venice*." *English Literary Renaissance* 24, no. 2 (1994): 372–85.
Strout, Nathaniel. "*As You Like It*, *Rosalynde*, and Mutuality." *Studies in English Literature* 41 (2001): 277–95.
Sullivan, Garret A., Jr. *The Drama of Landscape: Land, Property, and Social Relations in the Early Modern Age*. Stanford University Press, 1999.
———. *Memory and Forgetting in English Renaissance Drama: Shakespeare, Marlowe, Webster*. Cambridge University Press, 2005.
Summers, David. *The Judgment of Sense: Renaissance Naturalism and the Rise of Aesthetics*. Cambridge University Press, 1990.
Sutton, John. "Spongy Brains." In *Environment and Embodiment*, ed. Mary Floyd-Wilson and Garrett A. Sullivan Jr., 14–34. Palgrave, 2007.
Swarbrick, Steven. *The Environmental Unconscious: Ecological Poetics from Spenser to Milton*. University of Minnesota Press, 2023.
Swarbrick, Steven, and Karen Raber. "Introduction: Renaissance Posthumanism and Its Afterlives." *Criticism* 62, no. 3 (2020): 1–14.
Szondi, Peter. *An Essay on the Tragic*. Stanford University Press, 2002.
Tasso, Torquato. *The Liberation of Jerusalem*. Trans. Max Wickert. Oxford University Press, 2009.

Terada, Rei. *Feeling in Theory: Emotion After the Death of the Subject*. Harvard University Press, 2003.

Tinterow, Gary. "Raphael Replaced: The Triumph of Spanish Painting in France." In *Manet/Velázquez: The French Taste for Spanish Painting*, ed. Gary Tinterow and Genevieve Lacambre, 3–66. Yale University Press, 2003.

Traub, Valerie. *The Oxford Handbook of Shakespeare and Embodiment*. Oxford University Press, 2016.

Tribble, Evelyn. *Cognition in the Globe: Attention and Memory in Shakespeare's Theatre*. Palgrave, 2011.

Tribble, Evelyn, and John Sutton. "Cognitive Ecology as a Framework for Shakespearean Studies." In *Shakespeare and Emotion*, ed. Katherine A. Craik. Cambridge University Press, 2020.

Turner, Henry. "Worlds of Experience: Fiction, Image, and the Ontology of Ideas in Sidney and Shakespeare." *Studies in English Literature* 62, no. 1 (2022): 217–45.

Vasiliauskas, Emily. "*Antony and Cleopatra* and the Art of Dying." *Studies in Philology* 117, no. 3 (2020): 488–519.

Virgil, *Aeneid*. Trans. A. S. Kline. 2002. https://www.poetryintranslation.com/Admin/Copyright.php.

Viterbiensis, Joannes. *Liber de regimine civitatum*. Ed. Caietano Salvimini. In Aedibus Successorum Monti, 1901.

Volk, Mary Crawford. "On Velázquez and the Liberal Arts." *The Art Bulletin* 60 (1978): 69–86.

Watson, Robert N. *Back to Nature: The Green and the Real in the Late Renaissance*. University of Pennsylvania Press, 2006.

Weber, Samuel. "Drawing—the Single Trait: Toward a Politics of Singularity." In *Crediting God: Sovereignty and Religion in the Golden Age of Capitalism*, ed. Miguel Vatter. Fordham University Press, 2011.

West, William N. *As If: Essays in As You Like It*. Punctum, 2016.

Williams, Robert. "Italian Renaissance Art and the Systematicity of Representation." *Rinascimento* 43 (2004): 309–31.

Wilson, Luke. "Promissory Performances." *Renaissance Drama* 25 (1994): 59–87.

Wolfe, Cary. *What Is Posthumanism*. University of Minnesota Press, 2009.

Zajac, Paul Joseph. "The Politics of Contentment: Passions, Pastoral, and Community in Shakespeare's *As You Like It*." *Studies in Philology* 113, no. 2 (2016): 306–36.

Index

Christopher Pye is Class of 1924 Professor of English Emeritus at Williams College. He is the author of *The Storm at Sea: Political Aesthetics in the Time of Shakespeare* (2015), *The Vanishing: Shakespeare, The Subject and Early Modern Culture* (2000), and *The Regal Phantasm: Shakespeare and the Politics of Spectacle* (1990), and editor of *Political Aesthetics in the Era of Shakespeare* (2020).

www.ingramcontent.com/pod-product-compliance
Ingram Content Group UK Ltd.
Pitfield, Milton Keynes, MK11 3LW, UK
UKHW041846031026
15587UKWH00004B/1035

9 781531 513832